"With formidable wit and vivid powers of observation, Jennifer Jolly's style is a cross between Monty Python and Michael Palin. She writes about the perils of travel in up-country Uganda in the 1960s against a background of ever-increasing political unrest and a chronic shortage of money. Family life takes place in a tent in the wilds of nowhere. Pregnancy makes no difference. By day, her husband Cliff abandons her to go in search of baboons. By night, elephants and buffaloes come visiting. A must-read for those who love travel and adventure."

—Vernon Reynolds,
author of *The Chimpanzees of the Budongo Forest: Ecology, Behavior and Conservation*

THE ELUSIVE BABOON

THE ELUSIVE BABOON

THE ELUSIVE BABOON

A Ugandan Odyssey

Jennifer Jolly

Full Court Press
Englewood Cliffs, New Jersey

First Edition

Copyright © 2017 by Jennifer Jolly

All rights reserved. No part of this book may be reproduced or transmitted in any form or by any means electronic or mechanical, including by photocopying, by recording, or by any information storage and retrieval system, without the express permission of the author, except where permitted by law. The events in this book all occurred and, save for slips of memory, have been faithfully recorded. Some of the names have been changed for reasons of privacy

Published in the United States of America
by Full Court Press, 601 Palisade Avenue,
Englewood Cliffs, NJ 07632
fullcourtpressnj.com

ISBN 978-1-938812-99-6
Library of Congress Catalog No. 2017934099

*Editing and book design by Barry Sheinkopf for Bookshapers
(bookshapers.com)*

Cover art courtesy shutterstock.com

For my family, with love

Acknowledgments

I owe thanks to a number of people who saw some of the preliminary chapters and gave their helpful thoughts. These include the Writers of Metuchen and writers from Hoboken. A big note of gratitude goes to teacher, writer, and friend John McCaffrey. Without his encouragement, I would never have completed this book, and it would still be a pile of papers and unfinished stories on a shelf.

Jill Dearman, Isabel Urra, Cliff Jolly, and Caroline Jolly read and gave helpful comments on an early draft of the manuscript, while Cliff gave invaluable feedback about factual information on a later version. I also want to thank Barry Sheinkopf, who has so ably helped me through the final stages of editing and production.

Finally, I give grateful thanks to my family, Cliff Jolly, Caroline Jolly, and our son Erik, who now lives in Vermont with his wife Imogen and their twin sons, my very dear and only grandchildren, Liam and Owen. They have all always given me their kind and generous support.

The Republic of Uganda

PART ONE: BABOONS ON THE BRAIN

> "He who understands baboons
> would do more towards metaphysics than Locke."
> —*Charles Darwin:*
> *jotting in a notebook, 1838.*

PREFACE

In August 1971 we were readjusting to life in Manhattan. The humidity was high. Temperatures soared into the upper nineties. Glass-windowed buildings, concrete sidewalks, and the chrome on cars reflected the sun's blinding glare and added to the unbearable heat. From our air-conditioned thirteenth-floor apartment in Greenwich Village, I gazed out at skyscrapers. The Twin Towers of the World Trade Center were nearing completion. Along Bleecker Street, drunks and addicts slumped in shop doorways amid discarded bottles and old food wrappers. Refuse littered the dusty sidewalks and swirled in the gutters. Garbage bags dumped at the curbside emitted the putrid smell of rotting food. Fire engines clanged, police sirens wailed, cab drivers leaned on their horns and shouted abuse through open windows. New York was going through a period of social and economic decay, when many fled to the suburbs.

Everything was so different from the environment I had just left in Uganda, where I had been able to simply open the door and wander onto grass. I missed the green forests, the clear skies, the mountains, the lakes, the majestic animals, the beautiful birds and butterflies, and the clear night skies with millions of stars that twinkled in a velvety blackness.

We had gone to Uganda twice for my husband to unlock some of the mystery surrounding evolution by watching wild baboons, trapping them, taking their blood, and analyzing it using techniques totally new at the time. When we first went in 1965, I had been a naive young woman living in London. That time we traveled by sea to Mombasa and by steam-engine to the interior of Africa, taking our small daughter. Along the way, we escaped an attack in Genoa, and I discovered I was pregnant. The baby would have to be born in Africa, but I anticipated no problems. How mistaken I was.

Our base was on an old coffee estate near Kampala, the capital. We traveled around in a battered old Land Rover and camped with few amenities as we aimed to trap baboons while living through violent thunderstorms, and seeing dangerous animals and vultures waiting to pounce on rotting carcasses. But the countryside was beautiful and not yet touched by modernity.

I'd always seen myself as invincible, but this image was shattered when I was due to give birth and a violent civil war broke out. Mass killings took place, bodies were dumped in Lake Victoria, and people disappeared without a trace. It was dangerous on the roads. A curfew was imposed. During this mayhem, our son was born, and I was lucky to survive a complicated labor. I was much more fragile than I thought.

In 1967 we moved to New York and returned to Uganda in 1971 to study the eating habits and behavior of baboons. This time we took our two small children and had to adjust to living in an isolated old rest house with no amenities on the edge of the Budongo Forest. While there, rumors began to circulate about the Ugandan army killing people on nearby roads, and I experienced gut-wrenching panic from the fear of being beaten up or shot by guns poked through the window. Our car was unreliable, but we managed to visit the beautiful Queen Elizabeth Park and escape before Idi Amin, the "Butcher of Uganda," committed the worst of his atrocities.

In spite of all the turmoil we were in Uganda during an interesting period in its history. Three years before we arrived, it had gained independence from Britain after one hundred years as a protectorate and was still adjusting. And so in this book I relive my memories of traveling to Africa, how we adapted to a different culture, met colorful local characters, saw excitingly unfamiliar wild animals, survived several life-threatening situations, and learned the hazards of scientific field research with wily, unpredictable baboons. My story begins in London in the early Sixties, when primate field research was in its infancy and Jane Goodall first came to the public's attention.

CHAPTER 1

Baboon Man

A COLLECTION OF FOSSIL REMAINS OF EXTINCT, giant baboons was gathering dust in a remote corner of the British Museum of Natural History. Some had been collected at Olduvai Gorge in Tanzania, and at Olorgesailie and Kanjera in Kenya, by the renowned archaeologists and paleoanthropologists Mary and Louis Leakey. Some had been collected as early as the 1930s, and no one had studied them since.

It was now 1960, the year that marked the beginning of a decade of dramatic social, economic, and political change. That year John F. Kennedy won the closest presidential election of the century in the United States; the Soviets shot down a US U-2 spy plane and captured its pilot, Gary Powers; a furious Nikita Khrushchev pounded his shoe on a desk at the United Nations; Xerox introduced the paper copier machine; *Psycho* was the most talked about film; and birth control pills were approved by the FDA, opening the way to a sexual revolution. A little known English rock group with the strange name of *The Beatles* gave the first performance of their careers in Hamburg, Germany, and Britain's Prime Minister, Harold Macmillan, gave a significant "Winds of

Change" speech signaling Britain's intention to grant independence to a number of its African colonies.

Later that year, a young anthropologist, my future husband, Cliff Jolly, dusted off the baboon fossils, laid them on a desk, and pored over them. He was on a quest as he measured them and made notes on these one-to-three-million-year-old relics. These dusty fragments had once been living, breathing primates, sharing their African habitat with early human ancestors. Could their lifeways, interpreted from their bones and teeth, help to unravel some of the mystery surrounding human evolutionary origins? He would use the results as the basis for his Ph.D. dissertation.

But Cliff wasn't supposed to be there. In 1957, three years before Macmillan made his "Winds of Change" speech, Cliff had been accepted at Oxford as an undergraduate to read law, thinking he could go into the Colonial Service as a route to studying human cultural and physical evolution. But when he realized he could achieve his goal more directly by studying anthropology, he had turned down the place at Oxford, much to the chagrin of his headmaster, himself an Oxford man. His parents too were disappointed. Neither of them had been to college, and here was their only son saying no to one of the most prestigious universities in the world. But he had made up his mind. And his life set off on a totally different course.

The Anthropology Department at University College, London, readily accepted him as an undergraduate. The program focused on social anthropology, but also included prehistoric archaeology and physical anthropology—the study of blood, bones, genetics, and evolution. Cliff chose to focus on physical anthropology and, in doing so, made a complete switch from his previous high school studies of German, French, Latin, and history. A brilliant Latin scholar, he used his Latin to figure out the etymology of almost any word thrown at him.

I once asked, "What does 'solipsism' mean, Cliff?"

He looked up from reading: "Surely you know that, Jen? It's the theory that self is the only reality. Comes from *solus*—alone, and *ipse*—the self."

"Hmm," I said, somewhat miffed by the implied put-down, and went off to find a more obscure word. He seemed puzzled that others didn't think the way he did.

Although he had never formally studied the biological sciences, Cliff had always taken a keen interest in animals, birds, and plant life. An avid bird watcher with exceptional eyesight and hearing, he could readily spot the detailed markings on birds and identify their songs, and he knew the Latin names for many species of plants. A voracious reader with a retentive memory, he could talk to experts about fossils, birds, mammals, plants, and geology. He also had an extensive knowledge of English history and appreciated old buildings and prehistoric monuments. Physical anthropology turned out to be a good choice, and Cliff eventually became an acknowledged expert in that field. But I'm getting ahead of myself.

For some reason, this brainy man, who some said looked rather like the young Michael Caine without glasses, took a fancy to me, though we were opposites in many ways. He was intellectual, I was practical; he was logical, I was emotional; he looked at things in depth while I skimmed the surface. We had both been born just before World War II broke out, but he, like Darwin, was a creative Aquarius; I was a many-sided Gemini.

We came from different parts of England. Cliff was an only child who grew up near to London and all it offered in the way of museums, the zoo, theaters, art galleries, and music. I grew up with a younger brother and many cousins in an iron-and-steel town in North Lincolnshire, where the night skies lit up with the intense glare thrown off by white-hot molten slag tipped from huge ladles onto enormous gray solidified slag-heaps. Our town was industrial but situated in an agricultural county and surrounded by picturesque old villages. We could cycle to one named Alkborough, which had a rare and very ancient turf-maze named Julian's Bower, said to be either Roman or Medieval in origin. When you sat in that soothing place surrounded by quiet on top of a hill, you could look down onto the confluence of the dark gray-green waters of the Rivers Trent and Ouse, forming the River Humber, which flowed into the North Sea. It separated the East Riding of Yorkshire, to the north, from North Lincolnshire to the south. Tennyson's birthplace, in the small village of Somersby, was only forty-five miles away from us, and just eight miles away was Epworth, the birthplace of John Wesley, the Methodist preacher. Lincoln, with its magnificent mediaeval cathedral founded around 1088, was less than thirty miles along the A15, which followed the straight old Roman Road.

After the discovery of iron ore and the development of the steelworks in the mid-nineteenth century, our town had been formed when five villages, all with Anglo-Saxon and old Scandinavian names—Crosby, Ashby, Brumby, Frodingham, and Scunthorpe—amalgamated under the name Scunthorpe, though each village managed to retain some of its own identity. Scunthorpe thrived as a boom town during the 1950s, but people thought the name sounded funny and it became fodder for music hall jokes. Yet those of us who grew up there were attached to it. You could easily bicycle out of town into the surrounding countryside for picnics, people were friendly, and there was a strong sense of community. Our lives centered on the steelworks, the town's soccer team, cricket, horse racing, pubs, and families. Scunthorpe had a movie theater that we frequented as teenagers, and pantomimes like *Puss in Boots* were staged at Christmas, but if you wanted "real culture," such as Shakespeare, an opera, or art galleries, the nearest place was Leeds, about fifty-two miles away.

London was 180 miles away, a very long distance at that time, but in 1951, when I was twelve, I went there on my biggest school trip. We traveled by train and back in a day, setting off very early and returning very late, to visit the Festival of Britain in Battersea Park on the South Bank of the Thames. The Festival marked the one-hundred-year anniversary of the Crystal Palace, Joseph Paxton's wonderful iron-and-glass edifice, and aimed to lift the spirits of a nation still struggling to recover from the devastation of World War II. I was thrilled that my parents let me go, and excited to see the futuristic cigar-shaped

THE ELUSIVE BABOON

Skylon pointing to the sky, the mushroom-shaped Dome of Discovery, and souvenir shops where I bought a commemorative five-shilling piece. To me, London was a place of wonder, but it seemed almost beyond reach from North Lincolnshire, especially by car, because at that time the journey took a full day, as the road wound its way through the narrow streets of small country towns. Nevertheless, in my teens I set my heart on some day living in London.

In those days women were expected to marry by around the age of twenty-one and have their husbands take care of them. Women like me had three main career choices: secretary, nurse, or teacher. Those who wanted to be a secretary or nurse left school at sixteen. I was expected to go to university, which meant I would stay on until eighteen. I thought a career in teaching would be a reasonable choice, but my mother, herself a teacher, was adamantly opposed to it. Apparently she wanted something different for me, and her influence prevailed. I was not going to teach but I envied those who knew what their futures held because I had no clue. I was interested in medicine, because I thought I might be able to contribute in the neglected field of women's medicine, but wasn't sure I could handle life-and-death responsibilities. I was an excellent pianist but not at concert standards and, in any case, was nervous when I had to perform in public. I was a good tennis player, but that was no career. I had always had an interest in acting and liked the idea of being someone who was not me living in different worlds, but in high school I had been told that, at five-foot-nine, I was too tall. I shut up about it, became sick of people telling me how tall I was, and developed a huge chip on my shoulder about my height. My mother was six inches shorter and wore fashionable, expensive high-heeled Italian shoes to add to her height. I wished I could do that. My mum expected me to marry, but she also thought I should have a career and said ominously, "You never know what might happen." All I knew was that I was expected to go to university. . .but to do what?

As a teenager I was bored with school and only crammed at exam time:

Does well if she bothers to apply herself, read one of my reports. My mother threatened to take me out of school if I "didn't get down to it." I refused and did little but play tennis, hang out with friends, and listen to Elvis, Lonnie Donegan, Bill Haley and the Comets, and *The Goon Show*. My grades suffered, I experienced some dark days from depression, and I continued to remain unclear about my future.

I loved to daydream, watch the clouds, and wander in the bluebell woods. When introduced to Coleridge's "Kubla Kahn," I was mesmerized. A whole new world opened up with "caverns measureless to man." And then, in the midst of my emotional turmoil, I decided to study psychology, knuckled down to work (focusing on the biological sciences), and began to read about psychological experiments and world affairs.

I was delighted when I gained acceptance at Bedford College, London University, to read psychology. Finally I was going achieve my dream of going to London. I had just turned eighteen when I boarded the train to King's Cross and a new life in which I battled intense home-sickness. I was overwhelmed by London's size and isolation. I knew nobody and sorely missed my family and friends. Never before had I felt so lonely, and no one seemed to understand my northern accent. I just didn't fit in. But determined to stick it out, I began to settle down, appreciate all that London had to offer, and make new friends— including Cliff.

My opinionated, rather prudish, maternal grandmother had proclaimed in disgust that psychology was just "Freud and sex." She didn't want to hear me when I said I was interested in finding out about the physiological bases of behavior and the extent to which nature and nurture influence what we do. For me, psychology raised many interesting questions. Could psychological traumas etch physical grooves so deep in the brain that there was no way to fully erase them and the accompanying pain? What was the role of psychologists in industry, and how did they study the attention span of air-traffic controllers? How could stress from fatigue and noise affect behavior and memory? Why did people perceive things in such different ways? I had difficulty sorting through the ideas that crowded my brain, so would move onto the next question while Cliff was considering the first. "You've got the attention span of a grasshopper," he said in exasperation.

I married Cliff in 1961 and found most women I met had husbands or boyfriends in industry, law, dentistry, medicine, or some other "respectable" profession, while I confessed that my husband spent hours in the British Museum of Natural History, looking at fossilized baboon bones. To me this was new and interesting, but for many it was a conversation stopper, or worse was met by the question, "What the hell does he want to do that for?" to which I never had an adequate response.

Then I thought about the steelworkers from my hometown, in their flat working-men's cloth caps, and hunched over their pints of English beer, slaking their thirst in the pubs because their dangerous, hot, heavy work near the blast furnaces made them sweat profusely. Their hands were gnarled and darkened with ingrained soot; their fingers and nails were often stained dark-brown from smoking non-filtered Players, Senior Service, or Woodbine cigarettes. They played darts, bet on the football [soccer] pools and sometimes as well on the horses at Market Rasen, the track frequented by my dad, a local bakery manager, and his friends. If these people had been told someone had chosen to study baboon fossils as an occupation, they would have said in disbelief, "He must to be bloody barmy," meaning he must be crazy, or "Ee, by gum, 'e's a rum un," which meant "he's a very odd chap." Many southerners looked down on northerners like me because we "spoke funny." We in turn were suspicious of people from the south. They spoke differently, and we thought they didn't know how to do hard manual work. Cliff therefore had two strikes against him. He came from the south and was an academic with a strange interest in baboons. The only person who seemed intrigued by him was my mother, Thora, though I suspected she murmured on the quiet, "Whatever is Jennifer thinking about, to be interested in that baboon stuff? She always seemed so down to earth and normal before she went to London."

However, my mum liked the new and unusual, so she put a positive spin on Cliff's work by equating brilliance and oddness. "Look at those Sitwells and that strange Bloomsbury group. Brilliant, the lot of them, but odd! Not like us normal folk." Yet I knew she admired them.

To make matters worse, Cliff decided to grow a beard. Maybe it was to make him look older, to fit the image of anthropologist, or to imitate Darwin. I disliked it and had to agree with one forthright friend who said, "It looks like bum fluff." I couldn't understand why a good-looking man had chosen to grow a wispy beard like one of the *Three Billy Goats Gruff* and concluded that Cliff was thumbing his nose at the world. Like the prince who turned into a frog in a fairy tale, he was turning from clean-cut Latin scholar and potential lawyer to whiskery baboon man. I worried what others thought, but he didn't care. I watched his transformation and began to wonder whether he was turning into a different person from the one I married.

Meanwhile, primatology was gaining in significance. At that time, the person who arguably had the greatest influence on British Primatology was Dr. John Napier. Born in 1917, Napier was an anatomist and orthopedic surgeon who specialized in injuries to the hand sustained in World War II. He developed an intense interest in comparative functional anatomy, especially the evolution of the human hand, and this interest naturally extended to the hands of other primates, living and fossil. In 1964, with Louis Leakey and Phillip Tobias, he described and named *Homo habilis* ("handy man"). This early human ancestor was one of a cluster of fossil discoveries in East Africa that in the 1960s transformed notions of human evolution. Napier's enthusiasm, creativity, and generosity in sharing information had a major impact on his postgraduate students, many of whom went on to illustrious careers. Cliff studied with John and learned about the functional anatomy of baboons and other monkeys by dissecting their bones and muscles at the Royal Free Hospital School of Medicine. When we knew John he was in his early forties—a good-looking man who walked with a limp and carried a cane. I thought he cut a romantic figure and in my mind created an image of him in a colorful cravat, a swirling cloak, and a wide-brimmed felt hat like Toulouse-Lautrec's rendering of Aristide Bruant. Together with his wife, Pru, a leading expert on taxonomy in primates, John founded the unit of primatology at the Royal Free. This was the first center devoted to the study of non-human primates in Great Britain. He also established the Primate Society of Great Britain and publicized primatology through his talks on the BBC.

In 1963 Napier organized a symposium at the London Zoological Society in Regent's Park that was a milestone in the development of modern primatology. Cliff attended and said that Jane Goodall, the British primatologist, had caused a great stir when she reported for the first time that chimps used tools. Man was no longer the only tool-user. She also said these animals, like us, had emotions. Some scientists accused her of anthropomorphism when she talked about "living among chimps," gave them names, interacted with them, and described their social relationships, but Goodall ultimately had a profound and widespread impact because she captured the imagination of the public. Never before had anyone heard of a young woman heading off to Africa to live alone among wild primates and, to cap it all off, turn current thinking on its head.

This all coincided with Cliff's interest in primates. His detective work on the fossils gave some interesting indications of life in the past. Based on bones and teeth, he deduced they came from several species of baboons that had become extinct. Unlike later baboons, which were adapted to living in trees as well as on the ground, the fossil baboons were more specialized —adapted to walking on the ground in an area of open country grasslands. The fossils had been discovered in old lake beds, so the animals had probably lived near shallow lakes where seasonal flooding prevented the growth of trees but enriched the ground flora, especially the grasses, on which they fed. But why had they become extinct? That event probably coincided with a period of diversification of human cultures and the elaboration of hunting techniques. At both Olduvai and Olorgesailie, evidence of extensive hunting had been found, so baboons could have been a frequent prey during a period of occupation by early humans. Higher primates are also particularly vulnerable to hunting because their slow reproductive rate makes them dependent for survival on a high rate of success in rearing their young to maturity. Hunting may have tipped the balance against their survival. Whatever the cause, these baboons had died out.

Meanwhile, Cliff's focus shifted to modern baboons, especially the evolutionary origins of the many different species and populations, spread across most of Africa, in diverse habitats. Together with the head of Physical Anthropology at University College, Dr. Nigel Barnicot, he started to look at the genetics of blood proteins using techniques that were completely new at the time. They came up with the idea of trapping wild baboons, taking samples of their blood, and analyzing the genetic variation in protein composition to see if this gave clues to evolution. Cliff was to do the field work, and Uganda was chosen as the best place. Barnicot had a contact at Makerere University, in the capital of Kampala, which would provide an academic environment as a base; baboons lived in many locations, so there would be plenty to trap; and a lot of people spoke English in the former British Protectorate. But first they had to obtain funding, and Cliff needed to learn more about blood analysis.

CHAPTER 2

On Running Blood

ONE DAY I HEARD CLIFF TALKING to his friend Vernon, who studied chimps, about starch gel electrophoresis and running blood. I wondered what it all meant, especially running blood. What had this to do with baboons and their blood proteins? I listened to the discussion and asked some questions, but the answers didn't satisfy me; I wanted to see running blood in action. My mother would have said, "She always wants to know the far end of things," and she was right. I cooked up a scheme in which I would pay a surprise visit to Cliff's laboratory to see what running blood meant. I was not going to ask if I could go because I knew he would say, "Why do you want to do that, when I've told you all there is to know?" Privately I didn't agree, so I took time off and set forth one afternoon in my trendy white plastic trench coat and favorite black-and-white-checked John Lennon hat, feeling pleased with my idea. I hopped on the underground, emerged at Euston Station, made my way down Gower Street, and went through the entrance to the Anthropology and Anatomy Departments at University College. No one was around.

I slipped past the dark, wood-paneled walls of the Anatomy Department with its skeletons of long-dead gorillas and orangutans and smell of must and chemicals before I headed to the dimly lit stone stairs leading to the Anthropology Department. Maybe the skeletons were to blame, but an image of Jeremy Bentham suddenly popped into my head. That intellectual genius and child prodigy, who began to study Latin at the age of three and was later associated with the founding of University College, made a strange request during his lifetime. He asked that his body should be displayed in a wooden cabinet after his death. In 1832, his wishes were duly fulfilled, and he now sat in his cupboard with a glass front, his "Auto-Icon," in one of the college halls. His skeleton had been padded out, and he was dressed in his own clothes with a white cravat gathered down the front, a black frock-coat, fitted fawn trousers to below the calves, white socks, and black shoes. He had a large wide-brimmed yellowish hat on his wax head.

As an undergraduate, I had found Bentham's presence unnerving, thinking it very odd to keep a dead man on display, dressed in his hat and clothes from the 1800s. I also I had an irrational fear of death: To see a dead man's corpse sitting in a box and looking out with his glass eyes gave me the willies.

Starting up the dimly lit stairs, I thought of Jeremy in his cabinet and began to wonder what else might be hiding in the corridor recesses on each floor. It didn't help that I had recently been scared out of my wits by *Psycho*, with Anthony Perkins as the crazy murderer Norman Bates. Suddenly overtaken by fear, I galloped up the steps and, trembling, leaned against the wall to recover my breath and brace myself to race up the last flight to the lab. Having gotten that far to see running blood, I wasn't going back, but I was having second thoughts.

I scampered up the next flight. The third floor corridor was also dimly lit. Several wooden doors led off it, but they were closed and there was no sign of life. I looked around in a panic and spotted a door marked *Laboratory*. Without stopping to knock, and finding the door unlocked, I rushed in and found myself in a dusty, low ceilinged room piled high with books, test tubes, flasks of various kinds, and jars of chemicals.

I was glad to see my husband in the midst of it, his fair hair flopped over his forehead, his shirt sleeves rolled up to the elbows. He was poring over some gelatinous substance on a bench. A rack of test tubes containing blood stood to the side. I thought the only thing missing was a veil of white mist with a large-fanged creature rising from it. I was still thinking of dead Bentham and the more than creepy Norman Bates, but this reminded me of Frankenstein's experiments. As I gathered my breath, I realized I would have to explain my sudden appearance to Cliff, who was so wrapped up in his work he didn't say a word. I began to feel foolish and wished I had not come. I took off my John Lennon hat and, because I was feeling upset, resorted to a silly remark, "So who was the unlucky victim today?"

He made no snide comment but said distractedly, "Oh, we've just had a batch of blood sent over from the Lister; they're doing medical research on various monkeys. There's a lot for us to do."

He didn't elaborate, and I didn't press for more detail. He was concentrating on his task. I watched. Eventually he took a deep breath, put down a test tube, pushed back his hair with the back of his hand, raised his head, and asked, "What's got into you?"

"Nothing, why?"

"You burst in as though the devil was after you."

He hadn't asked why I was there without warning him first. So I told him about Bentham, Norman Bates, creatures lurking in the shadows, and how I'd flown up the stairs. A look of incredulity spread across his face. He couldn't understand why I should be frightened and imagine people and creatures about to spring out from hidden recesses in dim corridors.

"You know, Jen," he said, "you let your imagination run away with you. It's perfectly safe around here, yet you barge in out of the blue, talking of ghosts, dead people, and creatures lurking in corners while I'm in here minding my own business and getting on with my work." He shook his head. He didn't seem particularly annoyed, just bewildered.

I switched subjects. "I came here because I wanted to see running blood and starch gel electrophoresis."

"You talk as if they're people. I've told you, they're part of a process. We run blood to separate the components using starch gel electrophoresis. Got it?"

"Yes, but I wanted to *see* it."

"Whatever for? Anyway, why the heck didn't you ask me instead of sneaking in here and catching me by surprise?"

"Because you would have said no, and that you'd already told me."

"Humph," he grunted, but I noticed he didn't deny it. "Oh, well, as you're here, I'll show you. Then you have to leave—I've a lot to do."

"Good, that's what I want. I promise I won't stay long."

Without wasting words he explained that, once a sample of blood was obtained from an animal, the blood was put in a centrifuge to separate the serum from the red blood cells. Next, a small sample of the red blood cells, mostly hemoglobin was placed at one end of a piece of starch gel in an oblong glass dish. I estimated the gel was about six inches long, four inches wide, and half an inch thick. It was murky white and reminded me of the coating on jellied eels. Chopped eels buried in pale, slimy gray jelly had never appealed to me, but people like Cliff who'd grown up with them loved them.

Putting the image aside, I returned to the process and saw that the gel with the samples was positioned between two oblong tanks filled with a chemical solution. The tanks were wired up to an electric current. As the current flowed through the gel, the red blobs of hemoglobin samples began to move slowly along the gel. This was called "running the blood." As it ran, the mixture of proteins in the red cell samples separated into its various components, which could then be stained so that bands in different shades of blue appeared on the surface. The whole separation process for one gel took several hours. Ironically, running blood was very slow.

"What exactly has all this got to do with baboons?" I asked.

"All sorts of things."

"But what?" I was getting frustrated by the lack of information. He seemed to assume I knew, but when pressed he finally elaborated. "I don't have time to go into all the details, but it's been shown that humans vary in the number and kinds of blood proteins from one geographic area to another. We want to know if there are comparable differences among baboons and other primates. If there are, what is the significance?   Could differences be linked to evolution?   For instance, could there be different genetic variants in different environments? Could the information help in medical research? For example, are there certain blood elements that might protect against a disease specific to a particular area of the body? Would the analysis shed light on behavior? If we can do research among baboons in the wild, we might find some of the answers."

"That's interesting."

"Yes, so now you know the far end of it." He sounded like my mother.

"I suppose you're right."

He went back to work while I looked around. The laboratory walls were decorated with photographs of the finished process. They looked like small examples of Jean Arp abstract art. There were also photographs of chromosomes that looked like miniature pairs of elbow macaroni floating on a gray background. I started to compose a silly ditty to myself:

> *The blood is running in its gel*
> *That's shaped like cheese but doesn't smell.*
> *I see it's stained in shades of blue*
> *But what they mean I have no clue.*

"What are you muttering about to yourself, Jen?" Had Cliff, with his extra-sensitive ears, really heard something?

I was not aware of having said anything and certainly felt my nonsense didn't belong in that serious environment. I stared intently at the wall before I turned and replied, "Nothing, dear, not muttering at all. Just looking." He frowned slightly but decided to keep his thoughts to himself, shook his head, and returned to work. I knew he wanted to be left alone, but he also looked up and smiled, and then I knew he wasn't annoyed. "Well," I said, "I'm sorry I dropped in on you, but you must admit all this talk about running blood has a Dr. Jekyll and Mr. Hyde flavor. Pity the whole thing is so innocent; now I can't have any more gory fantasies."

He was already focusing on the work and I thought he was going to ignore me, but then he commented, "You imagine too much, Jen. It's perfectly harmless but quite serious stuff, you know. Now off you go. I've a lot to do. I'll see you down the stairs so you aren't attacked by wild creatures and ghosts of Jeremy." He left off what he was doing, saw me through the door, and waited at the top of the stairs. I skipped down, waved when I reached the bottom, and wandered off, thinking about those photographs and wondering what it all meant while making up more silly ditties like:

> *Was it gory blood I saw*
> *As I wandered through the door*
> *And pale white gel with no appeal*
> *Reminding me of jellied eel*
> *Or macaroni on the wall*
> *In grayish pieces very small?*

The trip had been worthwhile, for I better understood the process of running blood. How to interpret the results was another matter, but I realized that going to Uganda to collect baboons' blood and observe their behavior might provide some clues to the link between blood proteins, behavior, and evolution. I assumed that, if he was going, I too would go but wondered when this would take place and where I would fit into the picture.

CHAPTER 3

Gearing Up

TIME PASSED. IN 1962, ANDY WARHOL exhibited his iconic Campbell's soup cans, the idolized sex symbol and mentally fragile Marilyn Monroe was found dead at the age of thirty-six, Rachel Carson published Silent Spring, the first James Bond movie was produced, and Cliff continued to look for funds for Africa. I began to wonder whether they would ever come through. In February 1963, Sylvia Plath committed suicide around the corner from where we lived in Chalk Farm, London. By the end of that year Harold Macmillan had resigned as prime minister in the wake of the Profumo affair, when John Profumo, the Secretary of State for War, resigned after he lied in Parliament about a sex scandal in which he was involved. The Rolling Stones were gaining in popularity, Beatlemania was heading towards a shrieking crescendo, and I gave birth to our daughter, Caroline, twelve days before the fateful November day when President John F. Kennedy was assassinated. I had just returned from the hospital when footsteps rushed down the stairs, and our friends, Vernon and Frankie, who rented the top floor of the four-story house in which we rented the ground floor, burst in with the shocking news.

Kennedy was shot and killed. Nothing would be quite the same again.

In 1964, Muhammed Ali became the World Heavyweight Champion, Nelson Mandela was sentenced to life in prison, the United States passed the Civil Rights Act, the Beatles were constantly hitting the top of the charts, and I had almost given up hope of going to Africa.

Then Nigel Barnicot received a grant from the Medical Research Council and Cliff was given a year off from teaching beginning September 1965. I would take time from my job, and Caroline, who by then would be almost two, was coming with us to Uganda.

A lot was happening in 1965. We had moved into a house we shared with Vernon and Frankie in Camden Town, and in January had been in the crowd that paid respects at Winston Churchill's funeral. The Beatles performed in the first stadium concert in musical history when they played Shea Stadium in Queens, New York. On the radio, we heard reports of civil rights protests and the beginning of protests against the Vietnam War in the United States. In Britain, public opinion was against the war, and in response to public pressure, Britain's Labour Prime Minister, Harold Wilson, refused to send troops when President Lyndon Johnson requested them. This resulted in a hostile relationship between the two men. When Wilson wanted to fly to Washington to discuss the situation, Johnson was reputed to have remarked to an aide, "We got enough pollution around here already without Harold coming over with his fly open and his pecker hanging out, peeing all over me." So spoke the larger-than-life President about his medium-sized, pipe-smoking British counterpart.

Meanwhile, our focus was on going to Africa. Immunizations against smallpox, yellow fever, cholera, typhoid, typhus, and polio were mandatory, and we needed malaria pills. Cliff planned to capture animals, tranquilize them, and take blood samples before releasing them back into the wild. He needed syringes, bottles, and sedatives. The blood would be prepared in Uganda but analyzed in the lab in London using the running blood technique.

THE ELUSIVE BABOON

However, since blood perishes and breaks down easily, especially in a hot climate, it would have to be preserved and arrangements made for someone to collect it from Heathrow Airport after the samples were flown back on ice.

. Our plan was to fly to Uganda and send our equipment by sea, but then we heard stories of crates sitting for months on the docks at Mombasa, the destination for our own crates before their final voyage overland to the interior. We had only twelve months at our disposal and needed equipment as soon as we arrived, and so we decided against flying and booked passages on the Union Castle line, whose ships sailed through the Mediterranean Sea and the Suez Canal, and down the east coast of Africa to South Africa. That way we could travel with the crates, disembark at Mombasa, shepherd the crates through customs, ensure they were put on a train, and accompany them inland to Kampala. The journey would take around three weeks.

We also had to figure out the best clothing for watching and trapping baboons, trekking through the jungle, and living on the equator, none of which we had ever done before. After talking to people and reading books, we began to settle on our specific requirements, but not without some debate. Cliff said army clothing would be best. According to him, we needed the kind of gear Montgomery's men wore when they won the battle of El Alamein. I said I wanted loose-fitting, light-colored cotton garments—not flowing robes such as those worn by Lawrence of Arabia, but loose-fitting trousers and lightweight cotton tops. I also wanted a straw hat shaped like the one Katherine Hepburn had worn in *The African Queen*. The brim would provide shade. I would drape a colorful silk scarf across the top and tie it under my chin.

He uttered a short laugh. "Fashion is not the object here."

"I know, but I don't like the idea of army gear."

"I think it would be *perfect*. Let's go and see. We also need cooking equipment to use in camp."

I gave in. We left Caroline with a baby-sitter and set off.

In the 1960s, London had a number of army surplus stores that carried goods left over from World War II. They were popular among the younger generation, who especially liked the khaki cotton army jackets with buttons down the front, epaulets on the shoulders, collars that could be turned up to protect the neck, and lots of deep pockets. I had to confess the jackets would suit our needs, and reluctantly agreed the subdued colors would blend into the environment.

Cliff's face lit up at the thought of grubbing through old tin mess cans, compasses, hats, trousers, shirts, and other items from a previous era, especially if they were cheap and, in his eyes, a bargain. Army surplus stores were like a magnet to him. Inside, the goods were piled up in heaps on dusty wooden floors and on shelves that stretched to the ceiling. Woolen socks, jackets, shirts, slacks, shorts, and army hats, mostly in shades of khaki or with a green-and-brown camouflage pattern, were stacked together with gray aluminum cooking stoves and pots for camping. A smell of old clothes and mothballs hung in the air.

Cliff poked around and emerged every so often with a grin on his face, holding some object he wanted. I rested on a heap of clothing and said, "All right," to whatever he produced, because I soon got bored and wanted to leave.

He also looked for books and maps and any account of various terrains from the 1940s. Once he unearthed a small pocketbook entitled *Upcountry Swahili,* which he claimed was a valuable treasure and just what we needed to communicate in the field; it was added to the growing pile.

But searching for hours for tropical clothing, camping gear, and old books was not my idea of fun. It gave me stiff legs and a backache. I liked fashion and color. Cliff could have cared less about that. While he hated going into regular clothing shops, army surplus stores held little interest for me. In the end I had to be dragged in and he had to be dragged out.

Eventually we amassed a motley collection of New Zealand army jackets, several bush shirts, several pairs of long baggy trousers, some shorts, walking boots, canvas water carriers, khaki-colored metal food containers with *Rations*

THE ELUSIVE BABOON

For Six Men stamped on the side, camp beds, and sleeping bags. If khaki had been the color of the day instead of the shiny reds, blues, greens, yellows, blacks, whites, geometrical shapes, and plastics of Mary Quant, we would have been in the height of fashion—khaki from head to toe. Cliff assured me we had "the right things for the tropics." I had reservations.

We faced another problem. Most of the clothes were designed for men, and the shops had no rooms in which to try them on, nor could we return them. Apart from jackets, we had to make an educated guess as to size and fit. The outcome of this became painfully obvious when we tried on our baboon- hunting outfits a few days later.

I had been looking forward to wearing my African clothes, thinking I would be quite the ticket decked out in khaki trousers and bush shirt. After we put Caroline to bed and darkness was falling, we drew the blinds in the bedroom and switched on the lamps to provide a cozy glow. The bed under the eaves was covered in a hand-made quilt made up of hexagonal shapes in bright patterns of flowers, ovals, circles, and stripes that served to bring back memories of the clothes they had once been. A lamp with an apple-green luster base, and a pale-pink silk shade with outlines of purple peacocks, sat on an old Singer treadle sewing machine. A large chest of drawers made of rich, burnished mahogany bought on the Chalk Farm Road stood against one wall, and the wooden floor was brightened by a rag rug pegged together on sacking during World War II by my maternal grandmother. There was a faint smell of lily-of-the-valley perfume.

Cliff settled down on an old round Arab *pouffe* he had acquired in the Camden Town market during one of his many trips to the junk store. Strips of faded red, brown, and green leather fanned out from a central button on top, and its seams were beginning to split, causing shredded newspaper to spill out. Wondering where it had been in its past life, I began to imagine a room occupied by a fat pasha, lying on bright-colored yellow and red silk cushions, being entertained by exotic, long-nailed, heavily perfumed and bejeweled dancing-ladies in diaphanous skirts and veils. Then Cliff broke my reverie by clearing his throat, said he was all eyes, and told me to get going. He folded his arms and leaned forward with elbows on his knees.

I was wearing a tartan miniskirt of black and green wool, a lacy fitted black top, and apple-green shoes with chunky heels and black gross-grain bows from Russell and Bromley. I didn't have many clothes, but those I did were fashionable. I selected a pair of khaki trousers and a bush shirt and took off my skirt, top, and shoes.

I struggled into the trousers. The result was horrifying. They were about ten inches too large at the waist, too tight in the hips, and the legs finished well above my ankles. Next I tried the bush shirt. It turned out to have an enormously wide body and sleeves that barely reached my wrists, and looked as though it had been made for a short, fat wrestler. This was the result of my own lack of attention and of not being able to try things on. Even then I was amazed we had purchased garments so totally wrong in size. In minutes I had been transformed from vogue to frump. My dream of looking like Katherine Hepburn in *The African Queen* went out the window.

Grabbing a bundle of spare material at the waist of the trousers, I stared down at the creases where they lodged on the hips, pulled at the too-short sleeves, scowled, turned to Cliff, and told him I could not wear these garments. His mouth twitched as he tried not to laugh; then he composed himself, looked me up and down, and said he could see no problem. He conveniently ignored the tight fit at the hips and suggested I could draw the trousers in at the waist with a belt or even a bit of string. I could roll up the sleeves of the shirt, and they would be fine.

Flummoxed by this and not sure if he was joking, I reluctantly pulled on some tan-colored Spanish walking boots with thick crepe soles and fleece linings and laced them up. These huge clumpy things were a far cry from my lovely apple-green shoes. I said they would make my feet sweat. Cliff argued that boots were ideal. The lining would cushion my feet, and I would be glad of them when I trekked over rough ground and tripped over tree stumps. I retorted that I had no intention of tripping over tree stumps and thought tennis shoes would be far more sensible. By then I was upset by the horrible outfit and furious with him. I swore to myself I would never wear those clothes.

Meanwhile, he had tired of crouching on the *pouffe*. Pushing himself up, he unfolded his limbs, stretched to his six-foot standing height, said it was his turn, and proceeded to take off an orange-red shirt to which he was particularly attached. He wore it with a mustard-colored tie. It made me think of eggs sitting in tomato sauce. He exchanged his tomato shirt for a khaki shirt and khaki jacket before he removed his long trousers and put on a pair of khaki shorts. Finally he pulled on some army socks and boots. In this getup he took on the stance of a Spanish dancer, tossed his head, looked haughtily down his aquiline nose, and sniffed. "So," he said, as if bedecked in sartorial splendor, "what do you think of this fine New Zealand army jacket?"

I was still annoyed and ready to throw the Spanish boots at him. But it occurred to me that his long-awaited dream of going to Africa was about to materialize, and that he didn't want it spoiled by my complaints. In a mega-hit that year the Rolling Stones sang "I Can't Get No Satisfaction." That was how I felt as I gingerly lowered myself onto the *pouffe* in my army surplus pants and shirt that pulled in all the wrong places.

Meanwhile, Cliff patted himself down and tested the jacket pockets. I thought the body of it was too short, and the shorts, from which his pale English legs projected, were baggy, but reserved comment, asking only how they felt.

"Oh," he said, "the jacket feels fine. Maybe you could get a needle and thread to adjust the buttons, but the fit is good enough. Heavens alive, we can't expect Savile Row in the jungle. Just look at the pockets: Two on the chest, and two at the waist. Think how much I can put into them!"

I sighed as I imagined the notebook, pencils, bits of string, elastic bands, paper clips, test-tubes, keys, and numerous other small objects that would eventually be packed in there together with handfuls of gray fluff. I saw he was genuinely pleased with his purchases. "Yes," I said, "I suppose it really doesn't matter how things look if they're functional."

"Now," he replied, "you're talking some sense."

I refrained from further comment while he continued to walk around in his outfit. His boots, I noted bitterly, were not lined with heat-retaining fleece like mine.

Finally, we erected our new igloo tent in the middle of the room. Four hollow ribs led from four corners on the ground to curve up and over before meeting at a point on top. They had to be filled with air using a foot pump and a good deal of energy. Cliff got to work, and the tent began to rise at a ponderous rate, like a large behemoth bending and then straightening its knees as it heaved its bulk from the ground. Once it was up, we sat inside to admire the soothing green glow from the light that filtered through the canvas, and agreed it was ideal for our purposes.

The tent came down in a matter of seconds when you released a valve. A loud whoosh of expelled air followed, and the ribs began to bend and collapse, so that it looked like a large leaf-green elephant brought to its knees by the effects of a tranquilizing dart in its rear. After that, you had to jump on the ribs to try to remove the remaining air. Needless to say, it was impossible to get it back to its original pre-inflated size, and it refused to fit into the bag from which it had come. But at last we were ready to face the journey.

CHAPTER 4

Departure Day

WE WERE TO SET SAIL ON THE EVENING of September 30, 1965. Not knowing what lay ahead, I was apprehensive and excited when I woke, grabbed my old pale-blue woolen dressing gown, pulled it close to my body for warmth, and padded downstairs in bare feet to put the kettle on. I washed, dressed, fetched Caroline from her crib, gave her breakfast in her high chair, poured myself a cup of tea, sat down, and looked out the window. The sky was black. Rain was coming down in torrents.

London is known for wet days, but that autumn one was worse than usual. From the front window I saw the roads were shiny and slick. People in tan or navy-blue raincoats, with collars turned up and their shoulders hunched, carried umbrellas and briefcases on their way to work. The rain was bouncing off the sidewalks, beating against the windows, and splattering through the few remaining leaves on the trees. Tires swished through the wet surfaces on nearby Camden Road. A few lights twinkling in the windows of the concrete utility council flats across the square provided the only bright spot in the dreary scene.

Caroline chatted to her doll and teddy while I nibbled on a piece of toast and marmalade washed down by more hot tea, and thought how good it would be to get away from the chilly wet weather. I was absorbed in thought when Cliff appeared, grabbed something to eat, and lifted the remaining baggage onto the landing. Our very heavy gear had already been crated and sent to the George V docks, but the final lot of luggage had to be transported to the docks that day to be put on board ship with the crates. It included a metal cabin trunk.

Cliff's parents, and his Uncle George, were taking us to the docks in their two cars, but neither car could accommodate the trunk. Only the back of our small minivan could. We had decided that Cliff would drive the trunk with our other cases to the docks and drop everything off, ready to be loaded when the ship docked later in the day. He would drive back, leave the van for my brother to pick up, and we would all set off for the docks in the relatives' cars. We had to be on board ship by five at the latest, but the distance from Camden Town was only about nine miles. We believed there was plenty of time.

The relatives arrived mid-morning, shaking rain off their coats and hats, and commenting on the terrible weather. Then the men loaded the trunk into the tiny minivan, which was just big enough to accommodate it in its covered-in back. After the remaining cases were put on the passenger seat and floor, it was full. Cliff folded himself behind the wheel, checked the mirrors, started the engine, put the van into gear, revved up, waved cheerfully, went to the end of the square, and disappeared round the corner onto the Camden Road. It was almost eleven o'clock.

At noon we ate soup, ham sandwiches, cake, and hot coffee, thinking Cliff would be back by one at the latest. He could eat before we got into the cars and headed to the ship. But one o'clock came. He didn't appear. Then it was two. We needed to leave soon if we were to board on time, but the minutes ticked by, the rain continued to come down, and we began to look at each other uneasily, wondering what had happened.

THE ELUSIVE BABOON

At two forty-five, when he still hadn't appeared, I left Caroline with her grandparents and great uncle, ran downstairs, and looked towards Camden Road from the top of the front steps, where I was sheltered from the rain. Every so often I heard a car coming round the corner and my spirits lifted, but each time it was not our minivan.

I went back into the house, started to pace, chewed at my nails, and snapped at Cliff's mother when she asked if I wanted a cup of tea from a flask she had brought. Then I flung on my raincoat, grabbed an umbrella, headed outside to the corner of Rochester Square, and looked down Camden Road. Cliff was nowhere in sight. My mind filled with morbid thoughts. Maybe the van had broken down. Maybe he had had an accident. What would happen if we missed the ship and our crates went without us? There were no mobile phones. I had to wait, imagining the worst and becoming convinced we would never reach the ship in time.

By three-fifteen, when I had given up hope, I looked out the window once more and saw the van turn the corner. I ran downstairs, flung open the door, ran into the road, and yelled through the car window:

"Where the hell have you been? I've been worried stiff." I was almost in tears.

Under this provocation, he showed a good deal of restraint only saying through tight lips, "I've been stuck in endless traffic."

He grabbed the car keys, unfolded himself from the low seat of the van, stretched his legs, ran his fingers through his hair, and rubbed his eyes. His shoulders sagged. He hadn't eaten, but there was no time to stop. He would have to hit the bathroom and eat on the journey. Our immediate task was to get everyone into the two waiting cars and head to the docks.

For the academic year we had rented our flat to a young Canadian family and expected to have departed by the time they arrived, but we were so late they appeared with their four small children and found us still there. Chaos ensued as we prepared to move out and they tried to move in. People bumped into one another on the staircase, fell over bags, and rushed around getting in one another's way, but at last we were ready. We climbed into the cars, collapsed onto the seats, slammed the doors, and yelled our good-byes.

Cliff sat in front with his father, who drove; the women sat in the back; his uncle followed in his car. Cliff extracted a sandwich from a paper bag and explained through mouthfuls of food that he had been delayed by endless traffic jams. Instead of the estimated thirty minutes, he had taken over two hours to get to the docks.

On the dockside, a crowd of expensively dressed people in Burberry raincoats, cashmere scarves, and fine shoes was waiting under their umbrellas to board. Desperately short of time, Cliff drove our filthy little van into their midst, where he was greeted by rude comments in South African accents and people who waved fists when he hurled our scruffy bags into the middle of the Louis Vuitton suitcases. Trusting everything would be loaded on-board, he jumped back into the van and hit further traffic jams on the way back.

Now we all faced the journey. Again there were traffic jams. Horns honked to no avail. Anxious passengers looked out of taxi-cab windows as we inched forward. The rain continued to fall. I started to count the minutes and then the seconds. The atmosphere was tense. Cliff said little but kept looking at his watch. His father fussed like the White-Rabbit, saying, "Oh dear, oh dear." His mother was concerned but happy to have more time with her granddaughter. I folded my arms and hugged my stomach to still the butterflies and kept nattering in a pessimistic, unhelpful way. *We'll never get there. What happens if we don't catch the ship and our luggage goes without us?*

I didn't remember having to face such terrible traffic. When the docks finally came into view, it was past five, and we saw the gangplank of our ship being pulled up. We stopped at the dockside, jumped out of the car, waved, and shouted frantically. Thankfully they saw us and lowered the plank. We had no time to say a proper goodbye but grabbed Caroline, together with the remaining bags, and ran.

THE ELUSIVE BABOON

As I struggled up the gangplank in the rain, my new Vidal Sassoon hairstyle with short back and swinging sides was reduced to rat's tails. In one hand I was clutching various brown paper bags that were rapidly disintegrating in the downpour. In my other, I had Caroline. She clung to me, whimpering quietly like a small animal. Cliff's hair was plastered to his head as he brought up the rear with yet more small bags. Slipping and sliding, I made my way up the steep slope until we neared the top when, on wobbly legs with Cliff close behind, I took the final steps and almost fell onto the deck. The gangplank was immediately pulled up behind us.

From the ship's rail we looked down on the relatives in their fawn-colored mackintoshes and waved. Cliff's mother was wearing some clear plastic concoction, like a concertina, to protect her hair from the rain, and the men wore tweed caps. They looked up, gave a final wave, and turned around. The last we saw were their backs, bowed heads, and sad, sagging shoulders as they went back to their cars while we headed to our cabin to face the sea journey.

CHAPTER 5

The Dark Continent

I DREAMED I WAS SPEEDING along in a red car. White-knuckled, with eyes focused on the road ahead, I clung to the steering wheel and glanced at the sleek lines of the iconic silver jaguar on the car's bonnet. Gray, flinty tarmac rushed towards me and sped away in diverging lines under the tires. The bridge lay ahead. Patterns from its shining steel framework danced before my eyes as I raced towards it. The water below glinted and shimmered as orange-tipped waves caught the rays of the sinking sun. I had to cross the river by nightfall or be stranded in enemy territory. Figures with no definite form appeared in my peripheral vision and, as the car continued to gather speed, the noise of the engine grew louder, my attention began to slip, my body ached. Would I ever get there? Would it never end?

Suddenly a corner materialized, the car swerved and shot off the road, and I found myself spinning through the air in slow motion with nothing to hold onto, like Alice in Wonderland falling down the rabbit-hole. Just as I was about to land, I woke clutching at bed sheets and rigid with fright. But the noise of an engine continued. This was not my bed, this was not my room, there was nothing recognizable in sight, and I seemed to be alone. Totally disoriented in the unfamiliar surroundings, I struggled to get my bearings until the effects of the dream began to fade and memories of the previous day came flooding back. I recalled our last-minute rush to the ship, how we'd almost missed it after so many years of planning for our Ugandan trip, and the fear.

As my mind continued to clear and the tension lessened, I remembered I was in the cabin of an ocean liner. The noise was coming from the ship's engines. My husband lay asleep and out of view on the upper bunk. Our twenty-two-month-old daughter was beginning to stir in her crib to the side. With the vibrations of the ship all around, I rolled out of the unfamiliar bunk, pushed back my newly cut hair, crossed the room on bare tiptoes, headed towards the small round porthole, and looked out. Stretching ahead as far as I could see was a brackish expanse of gray-green water broken occasionally by bands of white foam. I was on the *SS Kenya Castle*, one of the ships on the Union Castle Line. It was October 1, 1965. We had departed the London docks in the night.

I tiptoed back, climbed onto the bunk, leaned against the pillow, and reflected on what I was doing in that small cabin in the economy section of an ocean liner. Years of preparation lay behind us; the unknown stretched ahead. I had no idea what to expect and had led a pretty conventional life. My favorite foods were fish and chips, roast beef and Yorkshire pudding, pork pies, and other English dishes. My previous travels had been limited to Europe, and my ideas about African food and customs shaped by my imagination and a few extremely limited childhood experiences.

I remembered three things about Africa from my childhood. One was a small object shaped like a circular African hut, about four to six inches high, that stood on top of the mahogany desk in my parents' dining room. The hut was light in weight and made of a substance somewhat like papier-mâché. It had a cone-shaped roof painted to look like yellow straw, which came almost to a point but instead fanned out like the head of a small sheaf of corn to form a flat circular top. A slot for inserting coins was cut into the slope of the roof. The walls were earth-brown and contained a small brown door that didn't open. On this the letters *C.M.S.* were painted in a creamy color.

"What does CMS stand for, Mum?"

"It stands for Church Missionary Society."

"What's that?"

"It's a society that collects money to help the people in Africa. The missionaries want to convert them to Christianity."

"Why?"

"They think the Africans should be Christians."

"Why's that?"

"Because the people out there are heathens, and the missionaries think they will be better off if they become Christians."

I didn't understand. Why was being a heathen bad?

"Do the people live in houses like that?"

"Something like that, I suppose."

"Why?" The hut was so different from our square red-brick house with its tiled roof.

She became impatient. "They just do. What do you want to know for? Why? Why? Why? That's you. Run along and stop asking questions."

I was about seven at the time, and the missionaries were not my main interest. That lay in pushing pennies through the slot in the roof and hearing the satisfying clunk as they fell into the dark interior. Unfortunately, I was unable to get them out. They could only be extracted when some brown paper stuck over a hole on the bottom of the hut was removed. Only the lady who came to empty the box was allowed to do that. She arrived at regular intervals during the year and sometimes came on Saturday morning when I was home from school. She would knock on the front door, open it—for it was never locked—pop her head through, and shout, "You-hoo! Anyone in? Are you there, Mrs. Brown?"

My mother would rush out from the kitchen or living room, adjusting her hair and glancing at her reflection in the hall mirror on the way. "Hello! Come in, Miss B. I've been expecting you."

She held the door wide while her visitor came through and waited to be ushered into the front room.

"Do you want coffee, Miss B., or maybe you'd like a cup of tea and a biscuit?"

"Oh, no thank you, Mrs. Brown. I have a lot of places to visit. Very kind of you, though."

"Well, let me know if you change your mind."

They would go into the dining room, and Mum would take the CMS box off the desk.

"Do sit down and make yourself comfortable, Miss B."

"Thank you, Mrs. Brown."

I watched the woman pull off her gloves and take a seat at the old Victorian mahogany dining table that had belonged to my maternal great-grandmother and great-grandfather Charnock. My mother joined her, and I waited impatiently for Miss B. to adjust the brim of her brown-felt hat, tuck back her gray-streaked brown hair, straighten her tweed skirt, and put down her bags before she slit open the bottom of the box.

"Keep your nose out the way, Jennifer. Why don't you run off and play?" said Mum as I tried to see what was going on. I didn't want to. I wanted to see what was in that box.

Miss B. emptied the large bronze English pennies onto the shiny surface of the table, counted them, gathered them up, put the money in her bag, and replaced the torn paper by sticking another paper over the hole before she turned to my mother and said, "Thank you very much, Mrs. Brown. The missionaries and poor people of Africa will be grateful." She rose from the table, pulled on her gloves, picked up her bag and case, and headed for the door. "See you next month, Mrs. Brown."

I wanted to know exactly where that money went and how it was used to convert heathens to Christians. What exactly were missionaries? What exactly were heathens? Where was Africa? Why were the houses so different from ours? No one seemed willing to tell me.

I'd shake the box to get an idea of how much money was inside but be told to leave it alone. Children were to be seen and not heard, and I was always in trouble for being inquisitive. In typical North of England fashion, where "our so and so" is used when referring to family members, my mother would say, "Keep your fingers off that, *our* Jennifer."

One of her favorite phrases was, "Curiosity killed the cat." I didn't know what she meant but didn't like the sound of it. She also said, "Little pigs have big ears," or "Who wants to know what you have to say?" or "Who asked you anyway, our Jennifer?" or "Why do you have to know the far end of things?"

Consequently, the missionary box aroused my curiosity but I had no satisfactory explanation of its purpose. Often I coped with the lack of information by going to an inner world of fantasies.

"Our Jennifer has a vivid imagination," said Mum as if this too was a bad thing.

At that time I knew little of the Victorian zeal to convert Africans to Christianity, and, when older, associated the Church Missionary Society with West Africa. Later I learned that members of the C.M.S. were partially responsible for arousing European interest in East Africa, when in 1844, the C.M.S. established headquarters on the coast in Mombasa. From there, Johann Krapf and Johannes Rebmann, Germans who had gone to live in England and joined the society, made the journey inland. In 1848, Rebmann saw the snowcapped peaks of Mount Kilimanjaro, the first European to do so. Krapf pushed further north and caught a glimpse of Mount Kenya in 1849. The reports they sent back of great lakes and snow-covered peaks in the midst of tropical Africa were met with much skepticism. Nevertheless, they inspired others to explore the interior further. When we arrived in Uganda, the impact of the missionaries still existed, especially in the schools and hospitals.

My second connection to Africa was my father's father, my Grandpa

Brown. In 1899 he volunteered and fought in the second Boer War, returning in 1901 as a young man in his early twenties. Grandpa was an old rascal who lived to be ninety-six and died without an enemy in the world because, as he told his wide-eyed grandchildren in his early nineties, he had "outlived all those buggers." The family would complain that he told stories *ad nauseum* about life in South Africa, but I later thought it was a great tragedy that no one had written them down. He was a great raconteur, especially after his evening tot of whisky, "the best medicine known to man." My dad often visited him in the evening after my grandma died, and he said that once Grandpa had his tot and refilled his pipe, he settled into his armchair to reminisce and, after Dad left, read his favorite author, Sir Walter Scott into the night.

Sadly, I heard hardly any of his stories, but remember visiting with my dad when Grandpa talked about his voyage between England and South Africa. He said people got sick on board ship, that many had died, and they'd had to throw the bodies overboard to bury them at sea. I was horrified to think of the sick, dead bodies going into the ocean. Grandpa was one of the lucky ones who had survived to tell the tale. He said there was a man named Shackleton on board. He turned out to be the same Shackleton who was later knighted to become Sir Ernest Shackleton after he became famous for his heroic exploits as a polar explorer in the Antarctic.

I had a memory of Grandpa saying Shackleton wrote a log of the sea journey, that he made copies on the ship's printing press, and one of them went to my grandfather, but after he returned to England, Grandpa lent it to someone and they never returned it. I had no proof of this except that my brother remembered being told the same story. It's certainly true that Shackleton was a Third Officer for the Union Castle shipping line, and served on the *Tintagel Castle*, a liner used as a troopship that sailed to Cape Town in 1899 carrying British troops for the Boer war, so I *was* sure Grandpa met him. I didn't know whether Shackleton wrote a record of the journey, but I wished I had heard more from Grandpa about the famous man.

My grandpa also met Winston Churchill when Churchill was a war correspondent in South Africa. Although a supporter of Churchill later in life, at that time Grandpa thought Churchill was a show-off and "full of himself." I wanted to know more about that, too. One of my cousins told me Grandpa said he had been at the siege of Mafeking when British soldiers were surrounded by Boer forces for months until finally saved. On one occasion he started to talk about the arrival of the Australian Mounted Rifle contingents, but as usual the family ignored him and spoke of other things, leaving me to wonder what he knew about Harry "The Breaker" Morant, the Australian horseman, poet, and convicted Boer War criminal who was executed amid much controversy in 1902.

Sadly, the only tangible evidence we had of Grandpa's service in South Africa was a silver pocket watch engraved *Mr. T. J. Brown of the South Staffordshire Regiment, as a volunteer in the war and given in remembrance of Queen Victoria on her birthday, May 24th, 1901*. There was also a photo at his house of Grandpa in his uniform wearing the distinctively brimmed hat that soldiers wore cocked up on one side. He was a small man, only about five feet four, who eventually sired six tall children whom he ruled with an iron fist.

Grandpa liked to tell stories about exotic, luscious African women he came across in his travels, but when he started to talk about their breasts and nakedness, the grown-ups would tell him to shush and, to my frustration, shunt my cousins and me into the front room to play. We could hear laughter coming from the back room while I was dying to know more about these women who were obviously so different from those around me. I knew naked breasts were not acceptable in England. Respectable women kept their nakedness to themselves, and talk about such things was considered risqué, but Grandpa loved to shock people. He would grin and show tobacco-stained teeth under his Charlie Chaplin moustache as he did so. Everyone commented how incorrigible he was, but this didn't stop him, and his children egged him on. I wondered whether we might see women like those Grandpa described.

THE ELUSIVE BABOON

The final connection to Africa came through one of my mother's college friends, Betty, who had gone with her husband to Sierra Leone in 1947 to teach for a year. When she returned, I was intrigued by the treasures she brought to our house. These included some round straw baskets with lids. The straw was arranged in circular bands, one on top of another, in brilliant yellows and reds, which seemed out of place among the pale creams and greens of our home. My mother decided one of them would make a useful sewing basket, and it ended up containing cottons, wools, and brightly colored buttons.

Betty also brought back two small snakeskin shoulder bags and a large crocodile handbag. My mother said the handbag was the right size to store pink and brown plastic hair curlers and brown hairnets, and so this exotic African bag ended up with mundane contents in a cupboard under the stairs. I had never seen anything like these African pieces in England, and they further sparked my curiosity about the so-called "Dark Continent."

To me Africa evoked images of primitive people, cannibals, witch doctors, and fascinating pygmies. I knew sophisticated societies and modern cities existed, though the beliefs and cultures of the people differed from ours, but I preferred to think of it as a continent shrouded by an aura of magic and mystery and was unable to comprehend its vastness.

At that time, I thought of the people from Africa as "natives" and had yet to learn to call them Africans. "Natives" was a commonly used term, but I learned it from my mother's Victorian mother, my grandmother Bickerton. Not only did she refer to Africans as "natives" but, like the people from the CMS, thought they were primitive, heathen savages who needed to be taught about God.

Even as an adult, nothing had fully prepared me for our trip, not even the colorful stories I had read about colonial East Africa by Elspeth Huxley and Isak Dinesen (*aka* Karen Blixen) or of the Mau Mau uprisings in Kenya. I had certainly never heard of people trapping baboons for scientific research of the kind Cliff proposed carrying out in Uganda. But it was time to stop day-dreaming, get out of bed, and see what the ship had to offer.

CHAPTER 6

A Narrow Escape

WE FACED THREE WEEKS AT SEA, and I feared life would be uncomfortable in our tiny economy-class cabin. All it contained was a metal washbasin, two single bunk beds one on top of the other, a child's cot, and a small wardrobe. After we brought in the luggage, there was hardly room to move. The porthole provided our only source of daylight. The toilet led off the hall outside. None of this fit my image of an ocean liner with luxurious wooden fittings, lots of space, soft beds, and waiter service.

Fortunately, the ship itself provided many amenities. On deck we discovered a library, a hairdresser, a barber, a gift shop, and a room that was converted to a dance floor at night and where films such as *Becket,* starring Peter O'Toole and Richard Burton, were shown during the day. A nursery with a trained staff was provided, but our daughter hated it. She wanted to trot around the deck to explore the life belts, ropes, and man-hole covers, watch the dogs onboard playing in the area allotted to them, watch people play deck games, and find new friends. Leading to the bowels of the ship was a short wooden staircase. We descended and emerged into a small area about six feet square lit by dim electric lights. To the left, an enormous refrigerator stocked with food for the journey hummed loudly and emitted an odd dank smell. Ahead of us, glass doors led to a dining area set up with small tables to seat six to eight, with the captain's table at the head.

During the three days from England across the unusually calm Bay of Biscay we quickly got to know people and ended up with a core group of six, all around the same age with children of similar ages. Ray and Barbara Barry, a tall, fair-haired friendly couple from Cornwall, were getting off at Aden. Emlyn and Barbara Jones, a couple from Wales, were going to Uganda, where he worked at Namulonge, an agricultural research station not far from our destination of Kampala. Many people on board were professionals who were working as chemists, doctors, biologists, or geologists in Africa. They had been home on leave and were returning for another tour of duty. Not surprisingly, we were the only baboon people.

On the fourth day, the distinctive, almost triangular shape of the Rock of Gibraltar began to appear ahead of us in the early morning. It jutted into the sea and stood way above the surrounding area. I thought it was forbidding and romantic, and liked the way the Phoenicians had named it one of the "Pillars of Hercules" because this captured the indomitable craggy white-gray limestone exterior that formed a fortress guarding the narrow passageway of the Straits before opening onto the Mediterranean Sea. Its impassive outward face masked an interior full of natural caves and man-made underground tunnels used for strategic defense during many battles. What stories it could have told from centuries past, if only it had a voice.

We moved slowly through the Straits into the Bay of Algeciras and dropped anchor. By then the weather was changing. The sun shone from a cloudless sky, and the waters were so calm they looked like a sparkling blue mirror. People discarded overcoats and wore lighter clothes. A convoy of small boats, each steered by a bare-headed, tanned, muscular seaman dressed in a white long-sleeved shirt and grey slacks, came across the bay in our direction. The men waved as they approached and pulled up alongside. Any passengers who wanted to visit Gibraltar for the short two hours of our stay could pay to be transported over.

People rushed to form a line and took turns climbing down into a boat. This would fill with about eight to ten people and then zoom off. When our turn came, Cliff climbed in before I handed Caroline to him and followed. Once full, the driver gunned the engine, which roared into action before the boat shot off across the clear water, sending up a white spray and creating waves in its wake. With hair streaming out behind, we hung on. On approaching the shore, we looked up. Sitting on the walls of the town and the rocks overlooking the water, we saw the famous Barbary apes with their grayish-brown pelts glistening in the sun. I had never seen such animals out of a cage before, and these were oddly named because they were not apes but macaque monkeys. Cliff said the species had been named "ape" when it was just another word for "monkey".

Our boat roared to a halt, pulled into the jetty, we clambered out, and, carrying Caroline, climbed up a path to get a closer view of the "apes." People were feeding them and snapping pictures while the animals sat impassively, nibbling on the food and quite unafraid of the curious visitors. Some people made faces to try to get a reaction, but the "apes" took little notice. They just stared and occasionally blinked their small, deep-set brown eyes. They seemed quite tame, but this was deceptive. If upset, they could inflict a nasty and dangerous bite with their sharp canine teeth. I was horrified to see people get close to them, apparently oblivious to any danger the animals could pose, and I held onto Caroline, who would have headed toward them with no sense of fear.

We climbed to the viewing platform at the top and could immediately see the rock's strategic military value. A commanding view opened up across the blue bay in which our own liner and numerous small ships were anchored in the untroubled water, and we could see for miles across the surrounding areas.

With time running out, we went down to the town and hopped into a cab. The place seemed desolate with little sense of community, although army people lived there. Sadly we had no time to visit the castle or the tunnels and caves within the rock. Mostly we saw souvenir shops geared to tourists. We also saw people on their way to work, lining up to go through English and Spanish customs in order to cross the land border of about three quarters of a mile between Gibraltar and Spain.

As soon as everyone was back on board, the ship set sail and moved steadily eastwards into the Mediterranean, going past the French Riviera towns of Nice, Saint Tropez, and the principality of Monaco jutting into the water. People began to wear sandals, shorts, cotton skirts or slacks, and light tops, as the weather grew warmer. The skies were clear, the air smelled fresh and slightly salty, and the grim, wet, cold gray weather of our London departure was becoming a distant memory.

At the busy Italian port of Genoa we docked to unload cargo, re-load and re-fuel, and, while there, had our first hair-raising adventure. An old friend, visiting on business, joined us the first evening for drinks. The following day was open. Cliff wanted to spend time working in the excellent Natural History Museum, and we agreed he should go in the afternoon. That morning we took a cab to see the Piazza Della Vittoria with its striking Arch to the Fallen dominating the center of the enormous square and built to commemorate Italian and Genovese soldiers killed in World War I. Then we visited street stalls piled high with fresh fruit or large chunks of newly caught fish that the locals bought.

In the afternoon, Cliff went to the Natural History Museum while I stayed on board with Caroline to watch a group of darkly tanned, muscular sailors and dock workers haul crates on and off our ship. Heavier items were picked up by large cranes. Men shouted loudly to one another through cigarettes that dangled from the sides of their mouths and stopped to chat, gesticulate, and show where things had to be placed. When women walked by, swinging their hips, the men stopped to watch, laughed, shouted, and bunched their lips in the form of a kiss or made pinching motions with their fingers: Some leaned forward to do just that, because it was common behavior among the Italian men.

By evening I was ready for a change of scenery. That afternoon, I had remembered our first brief visit to Genoa when we waited, with our friends Vernon and Frankie, for our small Citroen Deux Chevaux car to be loaded onto the ferry for Sardinia, where we spent a wonderful vacation because the island was still unspoiled and scarcely touched by tourists at the time. My recollections from that first visit to Genoa were of small wooden stalls near the dockside and cafes lining the narrow winding cobbled streets close to the sea front. A smell of saltwater mixed with fresh fish and the mouth-watering aroma of cooking with garlic and spices hung in the air. Men in clean aprons stood at the stalls, selling octopus tentacles and squid that were tender and delicious in their fragrant garlicky sauces. I remembered the port as quaint, and the people as welcoming, good-natured, and pleasant. I wanted to re-live the experience. When Cliff returned, I reminded him of it and suggested we should get someone to look after Caroline while we went on shore for a meal like the one we'd had before. He agreed.

That evening we prepared carefully, arranged for a baby sitter, fed Caroline, and put her to bed. I put on sandals, a flared blue cotton skirt, and a white cotton top to enjoy the warm Mediterranean weather, and Cliff changed into a clean white shirt and khaki slacks. We were ready for a good meal and, after checking once more on Caroline, set off. We soon came across stalls which, in my recollection, were our target. As we approached, a woman in a stained white apron, with her dark hair pulled back severely, stared at us out of black-brown eyes. There was none of the friendliness I remembered from before, but we ordered food and took it in packets to eat as we wandered along and enjoyed our surroundings.

I was starving and plunged my hand into the packet, extracted an octopus leg and popped it into my mouth, but when it hit my tongue a shiver of revulsion passed through me. It was rancid and tasted so bad that I spat it out. Cliff would normally eat anything and thought I was making a fuss. He took a very large mouthful and bit down. His face contorted into a grimace before he too spat everything out. He agreed the food was rotten.

Because we couldn't speak the language and the woman seemed hostile, we daren't go back to complain. We dumped everything and began to worry about getting food poisoning. I tried to convince myself the food was not dangerous because *someone* must be eating it and not getting sick. Meanwhile, I was trying to reconcile the experience with my memories from before.

Because we had been looking forward to our night out, we decided to find a small restaurant nearby and set off towards one of the streets I remembered as quaint, with attractive little shops and cafes. But my memory must have been playing tricks, because everything appeared totally different in the fading light. Maybe we were in a different area from the previous visit, but even so, I had thought it looked fine when we left the ship in the morning. Now smells of grease and dirt pervaded the air, and we found ourselves on a narrow cobbled street with tall old gray buildings rising on either side as we moved into a neighborhood that was poorly lit and contained sinister shadows. There were no places to eat, and in the sultry evening we came across women with black-rimmed eyes, red lipstick, heavily powdered faces, and wild beehive hairstyles in the fashion of the day. They lounged in doorways, smoked, and beckoned to men who walked past. Some solicited as they tottered along the cobblestones in their spiky high heels, taking small steps because their skirts were very short, incredibly tight, and clung to their protruding rear ends before going in at a sharp angle to their thighs, where they clung again. I watched in fascination, wondering how they got in and out of these garments. Maybe they were elasticized or they just pulled them up when at work. Many of the women looked unclean. Their clothes were not well washed, their hair was greasy, and their makeup piled on. I wondered if they were transmitting terrible diseases and shuddered.

It was time to get out, but that proved harder than expected. When we started to retrace our steps, the women began to take notice of us, especially when they heard us speaking English. They sidled up to Cliff saying, "*Allo, bebe,*" or, "*Allo, dahlin',*" as they rolled their dark eyes, flung back their heads, grinned in his face, and cackled. Cliff hunched his shoulders to protect himself and drew closer to me, thinking this would make them go away, but I was irrelevant. They were intent on their prey. Suddenly, one woman approached very closely and grabbed at him, causing him to draw back and utter in alarm, "They'll rip the shirt off my back if we don't get out of here smartly!"

He was right. At the start, I had been amused by his reaction, but now the atmosphere was charged and we were trapped among hostile strangers. The women closed in, began to taunt us, came close to Cliff, and began to poke at him in an aggressive manner, causing me to think of how people could gang up like a lynch mob, attack someone, and tear them apart. Things were getting too rough for safety. What if something happened to us while Caroline was back at the ship? At the thought of this, a wave of terror swept through me.

Cliff had had a nice peaceful afternoon contemplating monkey skulls and other inanimate objects at the museum, where he was troubled by no one and could sink into his own quiet, introverted world. I had had a relaxing afternoon and anticipated a pleasant evening. Now we were caught in a small space between tall buildings with only one way of escape. We had to go back the way we came; otherwise, we could get completely lost, and that could be fatal. I felt the hairs on the back of my neck begin to rise, my mouth became dry, and I had a sinking feeling in the pit of my stomach. I was no longer hungry but held onto Cliff and muttered, "We must keep moving," for now my only goal was to get back to our daughter.

I willed myself forward and refused to acknowledge anything serious could happen, but the women were not easily thrown off. They continued to pursue

THE ELUSIVE BABOON

us and grab at us for what seemed an eternity, though it may have been no more than ten minutes. All I could do was repeat, "Keep on walking, keep on walking, and please let us be safe." Finally we came to the end of the alley, saw our floating home, and the women stopped; we raced the last few hundred yards and scrambled on board. I had never been so glad to return to a safe haven. Cliff had been ruffled by the experience, but the only sign he gave of being perturbed was when he ran his fingers through his hair and said, "Oh, well, it could have been worse."

"Yes, we could be dead," I said bitterly, but we were both too shaken to say much.

I rushed to the cabin to check on our small daughter. She was sleeping peacefully. I was so relieved that I flopped down on my bunk and put my elbows on my knees and my head between my hands to quell the shaking and gather myself together before going to find food and drink. Cliff was with a group of people recounting what had happened. An older man who was listening said he knew Genoa well and told us the area we were in was not only the red-light district but also had a reputation for being dangerous. We had been lucky not to be hurt.

My emotions were all over the place when I knew our evening out could have had disastrous consequences. That night my sleep was broken by nightmares in which greasy faces framed with wild black hair stared at me, and hands started to grab me before I woke and lay breathing rapidly listening to the steady throb of the ship's engines. Our next stop would be Port Said and our first taste of Africa. I wondered what awaited us there.

CHAPTER 7

Arab Entertainers

ON REACHING PORT SAID, at the Mediterranean end of the Suez Canal, we anchored about thirty yards from the quay. We were spending a leisurely time by the pool when a commotion broke out on board as passengers rushed across the decks to crowd at the ship's rails. I jumped up from my pool-side chair, left Caroline with Cliff, and went to see what was happening.

A man made room for me, and I peered over the rail. Down below I saw the dark heads and brown shoulders of five or six skinny little boys bobbing in water that gleamed like black-green glass in the sun. They had swum out to the ship and were beckoning passengers to throw coins. A boy clapped his hands, shouted, "Eh, eh," and someone tossed a coin in his direction. It hit the water, sank, and the boy dived after it. A hush of anticipation fell over the crowd and a long time elapsed before the boy's head suddenly popped up like a cork from the surface of the sea. At this the crowd let out a collective "Ahh," like a sigh of relief, as if they had feared he would never come up again. The boy wiped his eyes, trod water for a few seconds to regain his breath, lifted up

THE ELUSIVE BABOON

his head and bared his teeth to show the coin held between them. He removed it, held it triumphantly in the air, grinned, and slipped it into a pocket of his shorts. Clapping broke out, rose to a crescendo, and subsided in anticipation of the next performance.

While this was going on, traders were rowing toward the ship in small, white boats. Each held two men dressed in gray trousers and white shirts that contrasted with their black hair and darkly tanned faces. One man rowed; the other stood and waved to our passengers and crew. A profusion of brightly colored goods was packed into each boat. I could see square and round leather *pouffes* with stripes down the sides, or a star-shaped pattern on top in various combinations of red, white, yellow, green, and orange. Cases of gold-colored necklaces, bracelets with bright-colored stones, thin bangles to stack on the wrist, dangling earrings, small metal pendants, and brass trays sparkled and shone in the sun. Leather bags, small carpets, and an assortment of children's toys were all wedged in. I could hardly wait to see them.

The traders pulled up to the side of the ship, secured their boats, climbed on board, and spread out across the decks to unload their colorful wares. Fascinated by the wealth of products, I went to look. Nothing attracted me, but curiosity and my weakness for jewelry took over. I went to look more carefully at the necklaces. Immediately, several men pounced and dangled gold-colored chains in front of me. They were far too gaudy for my taste, but each time I shook my head, the trader would lower the price, refusing to take "no" for an answer. And so I was introduced to bargaining as practiced in much of the non-Western world, where you are never expected unquestioningly to pay the first asking price as you would in England.

The traders spoke enough broken English for us to communicate and were extremely persistent. They kept lowering their prices, especially when I started to walk away, but eventually most of them gave up and went elsewhere—except for one. Even he began to give up in despair, but in the end won the game when he said, "You cruel lady. I have wife and children to feed."

That did it. I felt sorry for him and grudgingly bought a sun hat for Caroline, a small leather camel about six inches high, and a pair of shorts for Cliff, for which he had no use.

Several entertainers came on board. Of these the Gully-Gully man was most popular. He amused everyone with magic tricks centered on making chickens materialize out of hats and bags. Gully-Gully was thin and of an indeterminate age. He could have been anywhere from forty to seventy because his dark brown skin was wrinkled, parched, and weathered by the sun. Covering his thin frame down to his feet was a flowing, off-white cotton robe the Egyptians called a *galabiya*. It reminded me of the old-fashioned nightshirt worn by Alastair Sim in *Scrooge*. His round wire-framed glasses and shiny bald head made me think of Mahatma Gandhi.

Gully-Gully gathered a large crowd around him and started. First he surveyed his audience until his bright, bird-like brown eyes fixed on a man. Pointing at him with a slender, agile finger, he beckoned him into the ring. The man hung back until pushed forward by his friends, and then he entered into the spirit of the thing by jumping up and down and waving at the crowd, which responded with laughter and applause. Gully-Gully showed him a battered brown trilby hat, turned it upside down, invited the man to inspect the crown to make sure nothing was hidden there, and got him to agree the hat was completely empty.

Then motioning the man back to the crowd and thus deflecting their attention, he pointed at the hat, shouted, "Gully, gully, gully. No rabbit, no chicken, no mongoose!" slapped the hat several times, and an emaciated chicken flew out, screeching and flapping its wings. It must have been cleverly concealed in his tatty robe, and the crowd showed its appreciation with clapping and gusts of laughter. I had taken Caroline to watch, thinking she would be fascinated, but she clung to me in terror and hid behind my legs when the chicken appeared. After the show, Gully-Gully passed around a battered straw hat for donations, at which point many left. That seemed a bit mean to me, so

THE ELUSIVE BABOON

I gave him a coin but on reflection realized he had probably collected a reasonable sum and, if he caught all the vessels coming through, would make a decent living.

After the performers and traders had packed up and departed, the ship seemed quiet and dull. That night the ship passed through the first, long stretch of the canal, and we woke to find we had stopped in the Great Bitter Lake to wait for other ships to join us to go in convoy through the southern part of the Suez Canal. As the day wore on temperatures hovered around a hundred degrees Fahrenheit. The deck was hot underfoot, and with little escape from the relentless sun and no breeze to afford relief, people began to crowd into the pool to cool off. We waited for hours but finally moved forward around mid-day on October 13, about two weeks and two thousand miles after our departure from London. Growing up in England, I associated October with cold, damp, gray days that marked the beginning of winter. But on that October day, the sun beat down mercilessly, and the air was as dry as a bone. The heat grabbed and enveloped you, almost taking your breath away, and when we entered the Canal, I felt as though I was trapped in the hottest, most barren place on earth. I had never felt such heat, nor could I have imagined it.

On either side of us, yellow-white sands stretched interminably like a glaring mass of molten metal with the Arabian Desert to the east and the Egyptian Desert to the west. Mesas and dunes broke the landscape, and occasionally some palm trees marked small oases. Huge mirages shimmered and disappeared as we advanced, and I thought in awe of Lawrence of Arabia traveling over that barren burning-hot landscape and conquering its incomprehensible challenges. But I admired even more the Seventh Armored Division of the British Army, which saw distinguished service in the North African desert during World War II. Known as the Desert Rats, these valiant men hung on at the Battle of El Alamein and, against all odds, defeated Rommel's army as it advanced towards the Canal. Their extraordinary courage and fortitude contributed to a turning point in the war.

From the deck we saw the convoy of large ships following us. Many were oil tankers going to Arabian ports to load up before taking their cargo to Britain and Western Europe. Their huge shapes rose upward, and their width almost filled the canal. With desert extending on either side, they appeared to be sailing through sand.

The Canal stretched just over a hundred miles, and I'd read about the difficulties during its ten years of construction. Forced labor was used, and it was almost beyond belief to think that laborers in their tens of thousands, working by hand, had carried away so much desert sand in baskets. Due to the extreme aridity of the region they'd also had to dig out the Ismailia Canal, from the Nile near Cairo to Lake Timsah, to bring essential fresh drinking water. This added another seventy miles of hard work, but labor was cheap and life viewed as expendable. Sadly, thousands of laborers died during construction after suffering in inhumane conditions when they were probably forced to toil like ants until they dropped from hunger, thirst, and heat exhaustion. Needless to say, when it finally opened in 1869, to the delight of those in charge who apparently had no regard for the human suffering, the Canal had a dramatic and positive impact on trade from an economic and strategic standpoint. The British especially welcomed this "Highway to India" because it reduced the long voyage to their "jewel in the crown." Instead of sailing 10,800 nautical miles around the Cape of Good Hope, the distance from London to Bombay was now 6,200 nautical miles, a huge savings. The costs of transporting commodities such as tea, coffee and sugar were much reduced, and when the demand for oil increased during the twentieth century, the Canal's strategic importance became even more critical.

A few more days lay ahead. By now I was tired of this long journey and had started being nauseous all the time. I began to worry that something was seriously wrong.

CHAPTER 8

Sickness Strikes

ENGLISH PEOPLE TALKED INCESSANTLY about the weather. Will it rain today? Should we take an umbrella? Do we need a sweater? Isn't it miserable and cold? They moaned endlessly. If the thermometer hit seventy degrees Fahrenheit in the 1950s, we were in a heat wave. The weather was very changeable, but extremes of temperature were rare, and I never found English weather exhausting. Now the heat took all our energy as we sailed out of the Gulf of Suez and into the Red Sea.

Although the desert heat was intense, it had been dry and proved more tolerable than the heat and humidity we now encountered. Without air conditioning, our small cabin was unbearable. Our obliging steward found us an old sheet of cardboard to try to scoop in air through the porthole, but it made little difference. During the day, other parts of the ship offered some escape, but the nights gave no relief. I would look at Caroline's small form lying on her cot with her fair skin red all over and damp to the touch while her pale blond hair clung in wet strands to her skull. Seeing this, I was wracked with guilt for subjecting her to such conditions. Fortunately for us, she was a stoical little thing who withstood it all, and I don't remember her complaining. But the conditions were dire.

For almost three long days, we flopped about in this constant sauna, trying to remain as comfortable as possible, until we reached Aden and docked in the evening. Once more, Arab traders swarmed on board, but these men were selling cameras, radios, and watches at bargain prices. One passenger, smiling like the Cheshire Cat, told me he'd bought a hundred pounds' worth of equipment for forty-three.

Aden was still a British Protectorate but going through the uneasy period that preceded its independence. A state of emergency had been declared in December 1963 after a grenade was thrown at British officials at Aden Airport following the push for Pan-Arab Nationalism by Abdul Nasser, the President of Egypt. We dropped off some passengers, including our friends from Cornwall, and, although not forbidden to go ashore, were advised for our safety to remain on board. With Genoa still fresh in our minds, we stayed put. Those who got off said later there were police and machine guns everywhere, but there were no incidents. Hearing this, I wished we had gone to see for ourselves because we might never again have the opportunity.

After Aden, the stifling, humid heat lifted as we entered the breezy Indian Ocean. One day, we were thrilled to see hundreds of dolphins swimming parallel to our course, leaping out of the water with their sleek, wet silver-gray backs arced in the air and glistening in the sun.

Then there were the traditional celebrations on board when we crossed the Equator. These were presided over by a crew member dressed as Neptune, complete with beard and trident. People dressed up in fancy costumes, onlookers were thrown into the pool to the cheering of the crowd, and drinks were poured all around.

Fancy dress parades were held: One was for grown-ups, the other for children. I made a white mouse costume for Caroline with a tail created from rolled-up pieces of hand-towel and a body from toweling underwear. I drew whiskers on her face, grabbed her by the hand, and paraded her around the deck while she toddled reluctantly behind. To my delight she was awarded third prize in her parade, but I had made the mistake of thinking a child that age would be happy to be dressed up and was quite offended when she informed me in no uncertain terms that she hated it. She was particularly annoyed by her headgear, consisting of a pair of toweling underpants with cardboard ears sticking out of the legs. I thought they were creative; she didn't agree. Cliff turned away so she couldn't see him laugh. He obviously sympathized with her outburst, especially when she tore off her headgear in protest.

Meanwhile, I was suffering terribly from sickness. It started after we sailed from Genoa, when I began to feel queasy and throw up. For a start I thought it was due to the bad food in Genoa, because Cliff had also been sick, but it refused to go away. I emerged from the cabin to see part of the ceremony when we crossed the equator, and managed to get Caroline to the children's fancy dress parade, but for much of the time I had to leave her with Cliff while I stood in the cabin with my head over the basin throwing up. Cliff never suffered from any kind of travel sickness, but I had suffered from severe car sickness as a child and now the sea became my nemesis. I was thoroughly wretched and didn't care whether I lived or died.

I also began to realize my problem was compounded by morning sickness: I was in the very early stages of pregnancy. To start, I said nothing because I wanted to be sure I had made no mistake. I remembered how I'd pointed out before we left London that there was never a good time to have a baby, and that having one in Africa would probably be as good a time as any. But I'd been joking and had no intention of giving birth out there. So I was shocked when I found I was pregnant as soon as we set off, would be pregnant for the first eight months of our stay, and would have to give birth in Uganda.

I had mixed feelings. On the one hand, it would put a damper on my activities. On the other, I was glad to have another child and there would be about two and a half years between children, which seemed perfect. I also assumed that a second pregnancy would be easier than the first and could see no reason why having a baby in Uganda should be a problem. After all, women did it all the time and most without access to modern medical facilities. I was very cavalier about it. Never could I have anticipated how mistaken all of my beliefs would prove to be.

My sickness took over. If I heard the refrigerator humming and sniffed the stale air near the dining room, my stomach began to churn. Like Pavlov's dog, I began to associate the refrigerator with vomit and eventually found it impossible to go into the dining room. I daren't take any pills to counteract the sickness because the consequence of thalidomide pills given to counteract morning sickness, which resulted in deformed babies, was still very fresh in my mind. I was taking no risks.

For almost four more seemingly endless days we were at sea. The water was not especially choppy, but the constant motion of the ship, the greasy smells from the engine, and the stink from the area by the 'fridge were more than I could bear. Initially I had looked forward to the voyage, had eaten well and joined in with the drinks. Now I wanted the journey to be over and began to wish I had never heard of baboons and catching the wretched things. Nothing tempted me, and if I tried to force anything down, my stomach refused. I was getting dehydrated, starting to feel shaky, and wasn't sure I could endure much more. I began to look haggard. I lost more than ten pounds, a large amount for someone already quite thin. Therefore, my relief was palpable when we finally saw Fort Jesus, the Old Portuguese stronghold constructed from 1593 to 1596 to withstand attack from Arabs. It jutted into the sea and marked the large and busy port of Mombasa.

Exactly three weeks after we set off from London, we pulled into Mombasa's Kilindi Harbour. The gangplanks were lowered and, with unsteady legs, I walked down to step onto African soil for the first time. I was ready to cry with relief at being on firm land. Ahead lay the final stage by train to the interior.

CHAPTER 9

Into The Interior

AS THEY SWUNG HEAVY BAGGAGE from the ship to the quay-side, the muscled arms of African laborers glistened with sweat. We lingered briefly to watch them before making our way to a customs shed about the size of an airplane hangar. Inside we were greeted by the smell of sweating bodies and a hive of activity. Boxes, crates, and bags were piled up inside the vast, dusty space where, along with other bewildered travelers, we milled around to look for our possessions. African officials and porters shouted and gesticulated to one another, and I don't remember any orderly lines or directions. One young man stared in horror at the ground. His crate had been dropped from a height onto the unforgiving concrete floor, where it burst open, strewing his clothing, toiletries, pans, and other household equipment all over. The last I saw, he was on his hands and knees, trying to rescue his possessions.

We were lucky. Cliff's Uncle George was in the business of crating goods for shipping overseas and had made sure our belongings were securely crated and put on board in London. He had also arranged for an agent, Neil, to meet us in Mombasa. Neil arrived with his wife Joanna and, after we finally located our crates and cabin trunk, made sure everything was inspected, our paperwork in order, and everything was loaded onto the train to Kampala. Without these two friendly people to help us through the chaos, we would have been lost, and we saw first-hand how our crates could have been stuck in Mombasa for weeks. We had been right to travel with them.

But we had no time to explore Mombasa, with its long and fascinating history. For centuries Arab traders had exchanged iron implements for ivory, palm oil, rhinoceros horn, and tortoise shell, and in the eighth century a slave trade had grown up. Arabs and Persians had established trade links that reached as far as India and China. Vasco de Gama stumbled on Mombasa in 1498, and it was in the hands of the Portuguese until it was recaptured by the Arabs around 1700. British influence came relatively late when they leased Mombasa from the Sultan of Zanzibar from 1887 to 1963, at which point, just two years before we arrived, it had become part of the newly independent Republic of Kenya. African, Persian, Arab, Portuguese, and British influences all contributed to Mombasa's rich culture.

We wanted to visit the ancient and historic Fort Jesus and the spectacular beaches but the crates took precedence. They had to be loaded into the freight compartment of the steam-train to Kampala by October 22, ready for us to leave with them that evening. At six o'clock we climbed on board, waved goodbye to our new Welsh friends who were driving to Kampala, and promised to see them there. After that we walked down the corridor to our carriage and sat down on the seats in our *couchette*. Our journey was to take two nights and a full day.

The railway had enormous historical significance. Construction had begun in 1896, and I'd read that initially it moved rapidly, so that the rail reached Kisumu on the eastern side of Lake Victoria, about six hundred miles from Mombasa, by 1903. To lay a hundred miles of track a year was an astonishing feat, for the territory was extremely difficult, with heavy sustained gradients and many curves. You could see why the Africans called the railway the "Iron Snake." Tunnels and bridges had had to be built and it was hard to believe that everything had been achieved by hard labor without the benefit of modern machinery. The branch to Kampala that we would take was not completed until 1930.

I realized it was easy to take the railroad for granted and forget how hazardous and difficult construction had been when people rode on it in comfort. I'd read there had been an immense amount of human suffering, that hundreds of lives had been lost to disease and accident. One of the most disastrous and horrendous events gained international attention when construction workers were killed by a pair of maneless male lions during the building of a bridge over the Tsavo River in 1898. The lions stalked the compounds at night, dragged the mostly Indian workers from their tents, and devoured them. Estimates of people dead ranged from 28 to as high as 135. My blood curdled at the thought. Not surprisingly, this rampage terrified the workers, and at one point hundreds of them fled, causing a halt in construction.

Lieutenant Colonel James Henry Patterson, the engineer in charge of the bridge project, eventually shot the lions, and work resumed. He recounted his adventures in a 1907 best-seller, *The Man Eaters of Isavo*. He had the skins made into rugs, kept them on his floor for twenty-five years, and then sold them in poor condition to the Chicago Field Museum in 1924 for the then-considerable sum of $5,000. There they were mounted and put on display. I thought Patterson must have done well financially out of the whole episode, what with the mangy rugs and the best-seller.

Winston Churchill was a big proponent of the railway and, in his inimitably succinct, dramatic style, captured issues in its construction when he defended it in Parliament: "The British art of 'muddling through' is here seen in its finest expositions. Through everything, through the forests, through the ravines, through troops of marauding lions, through famine, through war, through five years of excoriating Parliamentary debate, muddled and marched the railway."

In Britain the construction of the railway caused heated debate. Those opposed questioned its value because of the hazards and expense involved and referred to it as "The Lunatic Express" or "Lunatic Rail." Those in favor said it would lead to economic prosperity, and they were right. Once finished, the British benefited a good deal because the railway opened up access to the interior and helped in their bid to colonize East Africa. Heavy equipment could now be transported with relative ease to commercial farming areas, and the cash crops of cotton, coffee, and tea exported. Before that the main transportation of goods was by ox-drawn wooden carts that carried comparatively light loads very slowly over short distances. The train also became popular with those who wanted to go on safari, including Theodore Roosevelt, who rode on it in 1909. It was a great way to travel.

Some extraordinary people helped build the railway. Many were Hindus, Muslims, and Sikhs brought in from British India as indentured laborers. They sailed courageously in small dhows and ships from Bombay and Karachi to land at Mombasa. The Sikhs were especially important, and it is likely that, without them, the railway would never have been completed. Many were already skilled craftsmen such as carpenters, blacksmiths, and masons. These extraordinary men rapidly learned the complex skills of fitters, turners, and boilermakers, all critical to the successful manufacture, assembly, functioning, and maintenance of the boiler that generated the high-pressure steam to drive the engine. In this kind of boiler, the tank is under so much pressure that there would be a major explosion if it burst.

Even after the railway was completed, the dedicated, hard-working Sikhs continued to take great pride in servicing the locomotives to keep them in good running order. Other railway Indians also remained in East Africa, where their descendants often became successful businessmen and shopkeepers. We came across many of them when we reached Uganda.

Suddenly, doors slammed and a whistle blew. We were off. The old steam locomotive began to puff, puff, puff, sending up small clouds of white smoke

THE ELUSIVE BABOON

from its funnel as it pulled out of the station and began to toil slowly along a rising gradient on the first leg to Nairobi. By then it was almost seven and, being close to the equator, nearly nightfall. In the dining car we found the tables covered with white cotton cloths and small vases of flowers. African waiters in white uniforms served us freshly cooked beef and vegetables, followed by mangoes. After that we returned to our carriage, prepared the bunks, put Caroline to bed and, feeling tired yet glad to be on the last lap of our journey, retired for the night well satisfied.

At sunrise the next morning, we were still south of Nairobi, crossing the Athi Plains of open grass and scattered acacia bushes bathed in golden light. There we were excited to see our first views of zebra, ostriches, Thompson's gazelles, and a giraffe. It was a beautiful scene. Drawing into Nairobi station, I thought of how Nairobi had grown up because of the railway, when it was known as Mile 329, and established as a supply depot to facilitate construction of the line into the Highlands. Over the years it had mushroomed into a large capital city, making it hard to believe that it had started as little more than a swampy, mosquito-ridden shanty town with tents set up for the workers.

After a short stay, we chugged out of the station and came across young children, dressed in raggedy clothes, jumping up and down at the side of the tracks. The girls wore cotton slips and the boys faded khaki shorts. They waved in excitement and smiled, showing gleaming white teeth against their dark skins, when we waved back. For a while they ran alongside, giving up only when the train pulled ahead of them, and we saw them disappearing into the distance before we began to chug north into the plains.

Mountains and volcanic peaks rose on either side of us when we entered the Rift Valley, and there were enormous expanses of brown-and-green land dotted with bushes but little sign of human habitation. Small white clouds drifted across the vast blue skies. About thirty-seven miles northwest of Nairobi, the cone of the extinct volcano Mount Longonot rose like a monolith to dominate the floor of the valley. A distinct dip at the top marked its crater.

We spotted Naivasha, the freshwater lake noted for its hippo population and many species of birds. Then in the distance we glimpsed the huge, strange, shallow soda lakes of Elementeita and Nakuru. They looked like huge patches of gray mist on the landscape and were said to have a high alkaline content and known for their variety of birdlife, with vast quantities of flamingoes on Nakuru.

Continuing on into the seemingly endless land mass, we began to get our first real feel for Africa. Coming from a small island nation, I was overwhelmed by the size and grandeur of everything and felt like a tiny ant in a gigantic bowl. Every so often we stopped to pick up and drop off passengers at small stations along the way and then heard the sound of unfamiliar tongues. Luggage banged as it was loaded on or taken off, doors slammed, and then we heard the guard's piercing whistle before we set off again.

And then we began to climb slowly into the Kenya Highlands, heading for Eldoret, where the track would turn west and make its way down into Uganda. Late that evening we pulled into the small station of Equator, where at nearly 9,000 feet above sea level, it was so cold we needed sweaters. The sounds of unfamiliar tongues drifted across the crisp, frosty air from the direction of the station. Stars in a velvety black sky, and small station lights, twinkled in the rather eerie blackness. The three of us were isolated, and civilization as we knew it seemed very remote.

During the night we crossed the border into Uganda and, on waking, were rewarded in the early morning with our first view of Lake Victoria, East Africa's inland sea, when we reached Jinja at the source of the Nile. With only about fifty more miles to go, the train chugged into a fertile area covered in banana plants with huge shiny green leaves. After more than three weeks of travel by sea and land we were in the country about which Churchill famously wrote in 1908: "Uganda is a fairy tale. You climb up a railway instead of a beanstalk, and at the end there is a wonderful new world. The scenery is different, the climate is different and most of all, the people are different from anything to

THE ELUSIVE BABOON

be seen anywhere in the whole range of Africa."

If his words still held true, we had much to look forward to in the months ahead.

CHAPTER 10

Finding Accommodation

WE WERE THRILLED WHEN WE FINALLY pulled into Kampala station in the early morning of October 24. Grabbing our bags, we picked up Caroline, headed to the door, and emerged onto a long, low platform at the end of which men were shouting to one another as they unloaded goods from the train. The 1930s station building formed a solid structure of brown-brick sandstone towards the middle of the platform. A profusion of brilliant red, orange, and purple bougainvillea spread over a white fence that bounded it. The sun shone from a cloudless sky, and we were enveloped in warm air. But something was amiss because, apart from the men unloading crates, there was no sign of life, and we had expected someone to meet us.

Cliff had corresponded with the head of the Anatomy Department at Mulago Medical School, Makerere University, before we left London. He had been assured that arrangements were being made for our accommodation, and that someone would welcome us at the train to take us to our new home. We put down our bags while he extracted a letter from the top pocket of his shirt, frowned, and glanced through it:

THE ELUSIVE BABOON

"Yes, it says we're expected today."

"Strange no one's here, isn't it?"

"Yes. There's no other train, so they must be meeting this one. Maybe they're a bit late or decided to stay in the waiting room."

Caroline was beginning to fret. Hot, tired, disappointed, and feeling irritated, I said, "Well you'd think they'd have seen us and come out. Are you sure you've got this right?"

He looked. "Yes, it says we're expected today. There's also an address on the letter. We can go to that if necessary."

"But I thought everything was arranged. You told me he said someone would meet us."

"Well, I understood they would. We wrote back and forth several times, and he said everything was fine."

"But why would someone *tell* us we would be met and at the same time give us an address to go to? It doesn't make sense."

I had had nothing to do with the arrangements and had not seen the letter, but it never had occurred to me there might be a problem. I glared at him. "So what do we do now?"

Worried and becoming exasperated by my questions, Cliff retorted, "Don't look at me like that. It's not my fault. Don't start getting upset. We'll have to work it out. We'll go to the station and see if anyone's in the waiting room."

But I *was* upset. We'd been traveling constantly for over three weeks. I'd been excited about being met and taken to our new home. Now I felt like a pricked balloon. We picked up our belongings and trudged to the sandstone building with little Caroline.

An elderly African was dozing in the ticket office with his head on his chest. He was the sole occupant and didn't look up. We waited, but no one appeared. I was ready to scream in frustration and fear. After some time, we realized we were waiting in vain. Cliff shook his head. "I don't know what's happened. Maybe we should see if we can get a taxi and head over to the address on the letter. Let's hope someone's there."

I was extremely alarmed by this turn of events. Suddenly everything was uncertain. I sighed deeply, took Caroline's small hand, and dragged myself outside after him. We were glad we had eaten an early breakfast on the train and were not hungry.

A rather battered-looking gray Peugeot labeled *Taxi* was parked at the curb a little distance away from the station. There was no driver, but Cliff said we should head towards it because the driver would probably appear. So off we went, and as we did so, the silence was broken by a crowd of African youths dressed in khaki shorts and faded shirts. They appeared as if by magic, gesticulating, waving, and shouting.

I grabbed Cliff. "What's going on?"

He stopped. "No idea."

Although they didn't seem hostile, I was frightened when the men drew closer and tried to seize our bags. We hung on tightly. The last thing we needed was to lose our luggage. We had no idea what was happening and what to do. Apparently, none of the men spoke English, we didn't speak their language, and we were the only white people around. Cliff imperiously raised an arm and tried to shoo the crowd away. For a moment the men hesitated before talking broke out again in their unfamiliar tongue.

"Maybe we'll be safe if we get into the taxi until the driver comes," I said.

At the word "taxi," another roar went up, and the men began to hem us in.

I shouted one of the few Swahili words I knew, "*Hapana,*" meaning "No." Nonplussed, they stopped for a minute, began to look at one another with puzzled frowns on their faces, and pointed at us and themselves before talking broke out again. We took this opportunity to edge forward. Cliff looked at the crowd, glanced at the taxi, and motioned me towards it.

"Are you ready? We'll jump in and take cover."

Another roar went up. The shouting increased, and black hands started to grab at us from all directions saying, "*Me.* Taxi, sir."

THE ELUSIVE BABOON

Suddenly we realized they were arguing about who was going to carry our bags and get the fare for the ride. We had no idea how poor so many of them were, nor did we realize they viewed white people as a source of good money. They wanted us to select one of them to drive us to our destination. The question was, how? Cliff said he thought it was best to avoid making a choice because the crowd might turn on the driver we selected, start a fight, or engage in more wrangling. He thought we should get into the taxi and leave it to them to sort things out among themselves. That way we could sidestep the issue. I had no other suggestion.

He leaned down and picked up Caroline. Then we edged forward through the crowd, reached the taxi, grabbed the door handle, opened the door, hopped into the back seat with our bags and small daughter, and slammed the door. Another roar went up, but then a skinny young man broke away from the crowd, leaped into the driver's seat, turned the starter key that was already in the ignition, gunned the engine, and set off with a screech of tires. We clung on, with me holding Caroline tightly. The remaining crowd scattered and shook their fists at the driver. Cliff recovered his balance, leaned forward, and read out the address. As he did, so the driver turned to look at us and took his eyes off the road.

"Makerere. Yes, sir!"

At this, the vehicle swerved violently towards the edge of the road before he righted it and continued at a dizzying speed. Along the way, he hailed people and wove from side to side of the road, with the car kicking up red dust while we were flung around in an alarming manner. Cliff was rarely perturbed, but his eyes looked as if they were about to pop out of his head. Stunned into silence, we clung on. Fortunately, there was little else on the road, because the driver seemed more interested in turning around and waving to his friends than in looking to see where he was going. He reminded me of Toad in *The Wind in the Willows,* who got behind the steering wheel of a car and had the time of his life racing along, shouting at friends and going "poop, poop, poop," until he crashed and landed in a mangled heap. I prayed this would not happen to us.

Suddenly the driver swung the wheel, swerved round a corner, tires screeching, and headed up a hill to the University campus with its roads bounded by trees and shrubs. Near the brow of the hill, we went through some iron gates and entered the grounds of Makerere University College. Then we shot over a maze of roads until we skidded to a halt and struggled out with our bags, packages, and child while the driver waited for payment. Cliff fished in his pocket for money and asked, "How much?"

The driver gave a number that seemed high for a fairly short ride, but not knowing any better we paid up. A huge smile spread across the driver's face. "Thank you, sir," he said, hopped back in his car, and immediately headed off. Later we found out you never paid the initial amount quoted but bargained. We had grossly overpaid. No wonder the man had looked so pleased and why there had been so much competition to convey us in the taxi. Others had probably been caught the same way.

As the sound of the engine faded, we found ourselves in front of a wooden door. Cliff knocked. Nothing happened. He rapped again. If no one answered, we had no idea what to do.

As we were considering our options, we heard welcome footsteps. The door cracked open, and a very large, sleepy-looking man, with tousled dark hair and wearing a red T-shirt and tartan shorts, peered out. It was Sunday morning, and he was having a lie-in. His eyes opened in amazement when he saw that three bedraggled strangers had materialized on his doorstep. Cliff told him who we were, and he politely invited us inside.

We had expected someone with a British accent, but this was an American who introduced himself as Glenn Russell. We sat down, explained our situation, and it soon became apparent we were not expected. Glenn told us that the head of the Anatomy Department had left the country a few weeks prior to our arrival to take another position and was not reachable. We gathered the event had been rather sudden, because no one had yet been appointed to take his place. In the meantime, Glenn was acting as head of the department.

THE ELUSIVE BABOON

He had heard about our proposed visit from Alan Walker, who had recently arrived from England. Cliff knew Alan from studying with him under John Napier. But Glenn said no one knew exactly when we were coming, and if it had not been for Alan, no one would have known a thing about us. He assured us that of course we would have been met at the station if we had been expected. Then he added that, as far as he knew, no accommodation had been arranged either, and that he didn't know what to do with us.

At this terrible and unexpected news, my excitement over settling in turned to fear and dismay. We were four thousand miles from home with very little money, a small child, and nowhere to go. We had planned the trip to Uganda for years, had spent over three weeks traveling out there; we wondered what had happened to the correspondence Cliff had had with the head of the Anatomy Department before we left England. He had written for permission to work in the department and been told it had been granted. He had written to say we were bringing a small child and must be sure there would be accommodation. He had been assured everything had been taken care of. None of this had happened.

Unfortunately the former head had neither given our correspondence to Glenn nor told anyone else about us. The biggest stroke of luck was that Glenn was actually at the address we had in our possession; otherwise we would have been stranded. Nothing added up. Why would someone send letters of assurance that all was taken care of and do nothing? Why had we been given Glenn's address without him being informed? There was no way to find out.

Still, we were greatly relieved when Glenn turned out to be sympathetic and promised to help.

But it was Sunday morning, and the college offices were closed. Glenn suggested we should go to a hotel for the night and see what could be done in the morning. Despondently, we picked up our bags and child, piled into Glenn's small red Volkswagen, and headed to the Speke Hotel in Kampala. Glenn promised to contact Alan, saying he might be able to help, and left.

After we settled into our room, we wandered down to the lounge to rest in comfortable armchairs covered in a floral pattern so typical of the Sanderson prints seen in English homes. African waiters flitted around in formal attire—white coats and black trousers. Wonderful fresh vegetables and delicious beef were served at dinner, and the place exuded a sense of calm. We began to feel cautiously optimistic about our plight. There *must* have been a mistake, the University *must* have been told we were coming, and things would be put right.

But our hopes were dashed the next day when the University officials said they knew nothing about us. They even became irritated when shown the letters from the departed head of Anatomy stating we were expected. We never discovered why.

However, there was one piece of good news. A house owned by the University was available about seven miles out of town on Katalemwa, an old coffee estate where faculty members lived at subsidized rents. Although the officials were willing to let us move in, they insisted we had to pay the full rent. This was fifty English pounds a month and happened to be the exact amount of grant money we had to live on.

Our housing expenses in London had been about sixteen pounds a month. Makerere wanted three times as much. We couldn't afford it. But we were so desperate that Cliff and members of the Anatomy Department continued to negotiate with the University. Cliff offered to teach if that helped, though it was not part of the original agreement. They were not interested.

While I waited with Caroline the negotiations continued until finally a compromise was reached. We would not be given the same concessions as other faculty members, but the rent was lowered slightly. With no other alternative, we accepted. Our budget would be extremely tight, and we would have to live very frugally but could drop all thoughts of an immediate return to England. We didn't have a vehicle to transport us to and from Kampala, but Alan said people would help until we had our own wheels. And so we headed to Katalemwa.

THE ELUSIVE BABOON

CHAPTER 11

Katalemwa

A HECTIC WEEK FOLLOWED as we moved to our new home, Cliff prepared to fly to a conference in Texas, the crates arrived, and I was inundated by an enthusiastic mob of young Africans begging for work. Young men wanted work as houseboys or as *shamba* boys who took care of the garden. "Boy" seemed a term left over from colonial times. Young women wanted to look after Caroline as *ayahs*. All of this was new me to me, and I had to learn to handle it. My main problem was that, even if I wanted to hire them, I had no money to pay anyone and had to keep turning them away. Meanwhile, we depended on others to transport us the seven miles into Kampala to buy food or get to Makerere. Cliff planned to purchase a vehicle once he returned from the States. Fortunately, Alan and his wife, Rikki, happened to live down the road. They were a godsend and gave us a tremendous amount of help.

Our house was more than we could have hoped for. Built as a bungalow, one of about twenty, it nestled like all the other houses on the estate in its own grounds. It was much larger than our two-bedroom duplex in London and far bigger than we needed, but we enjoyed the extra space with three bedrooms, a large living room, a dining room, a large study or fourth bedroom off the living room, a bathroom, and a kitchen. French windows led from the living room to a concrete verandah in front. Furniture was provided, and although sparse and spartan, served our needs. Even when hot outside, the indoor temperature was always comfortable due to the polished concrete floors, the long sloping roof extending over the verandah, the shade trees surrounding the house, and windows positioned so that, when opened, they caught any cross-breeze.

A large garden surrounded the house and opened to a vast expanse of coarse broad-leaved grass in front. This sloped down to a hedge, on the other side of which a gray tarmac road wound through the estate. In the distance, a valley with steep slopes of a lush dull-green and reddish-brown led up to the blue-gray horizon. Small plumes of white smoke spiraled upwards from that area, indicating that people lived there in houses hidden among the surrounding paw-paw (papaya) trees and banana plantations. Some evenings we heard the echo of party drums, and some nights were alarmed by the unearthly screams of a hyrax.

Wild raspberries grew in our hedge, together with red, orange, white, and purple bougainvillea. We had a hibiscus with salmon-pink flowers and bright yellow stamens, an acacia with feathery leaves, a night-flowering jasmine, two frangipani bushes, one with pink and the other with white waxy-looking flowers shaped like small propellers, and a passion fruit vine with round, purple-skinned fruit.

At the edge of the drive leading to the garage, the bare trunk of a paw-paw tree rose about twenty feet into the air, where it was topped by a small clump of leaves and yellowish-green fruits shaped like rugby footballs. I was amazed to see an African shin up it easily to collect the fruit. Two small concrete rooms, the servant's quarters, adjoined the garage. In front of them, a narrow path was bounded by a latticed concrete wall about six feet high to provide privacy.

At night we heard the noise from insects with the click of crickets accompanied by the croaking of small frogs and toads. Sweet, exotic smells from plants like the frangipani and mimosa lingered in the sultry air, and bats flitted through the garden, diving for insects. We were on the equator, so every day of the year the sun rose at six in the morning and darkness fell at seven in the evening. There were no seasons as we knew them in England. Between five and seven the heat abated, the air felt soft and warm, and people pottered in their gardens or walked down the winding tarmac road to greet neighbors before dinner. Everyone seemed very friendly except for one of our immediate neighbors. He kept a fierce dog that growled ferociously to keep people off his property. We heard he had fought in the Mau-Mau uprisings in the 1950s, and that his two-year-old son had later been killed in a tragic accident. The story was that the father had been backing up his vehicle, hit a bump and, not knowing his son was outside, accelerated to go over it. He discovered too late that he had run over his own child. No wonder he mostly he kept to himself. How could anyone come to terms with such an event?

Time passed, and we kept insisting we couldn't afford to hire any help. Africans who worked full-time were paid about 140 shillings a month, which at that time was about seven English pounds or sixteen American dollars. It wasn't an enormous amount, but as the total we had to live on was only fifty pounds for everything including rent, it was too much.

But people continued to beg for work, and we reassessed our situation. Our finances were stretched so thin that we could not employ the usual full-time houseboy and garden boy, but decided we could afford one of each on a part-time basis. Thus we ended up hiring John to clean the house, and Fred to take care of the garden. They both worked in the mornings. A *shamba* boy turned out to be essential, because we were expected to keep the garden in order, and our grass was so extensive it required attention on a daily basis. We could also offer a place to live, because the servants' quarters were included as part of our rent. In addition, we had to provide the boys with sugar and cooking oil each week. I would do all of my own cooking and look after Caroline.

In contrast to our arrangement, most people had a full-time houseboy who would typically be on call from about seven in the morning until six or later in the evening, though he would generally rest in the afternoon. Some of the houseboys had been trained to cook so well they could produce delicious Yorkshire puddings, soufflés, trifles, and other English dishes, which they themselves never ate. "Boys" who cooked like this, were generally older men who had been in service for years before independence and had spent a good deal of time with the same family, but few Africans were so highly trained where we lived. Most just wanted to earn enough money to help pay for the education that would allow them to look for a better job.

Jobs in which Africans worked for a household were highly sought after. They were given a home on the premises and extra provisions, and were able to grow most of their own food or food was given by their relatives. Their pay was a most welcome bonus.

One American woman steadfastly refused to hire any Africans, saying this was equivalent to slavery. I assumed she was particularly sensitive to the plight of Black Americans because the Civil Rights Act had been passed only the previous year. But this situation was different. I too had reservations about whether it was right to have servants and was adamantly opposed to slavery, but other people on the estate told me we were doing the Africans a great disservice by not employing them. The local Africans needed the money to help them support their families and their education.

When I realized how little people paid for help compared to their household incomes, I could understand why the English women in Uganda had so many children compared with those back home. A woman whose husband was the breadwinner on an ex-pat salary could easily have a full-time houseboy, garden boy, and *ayah*. Relieved of all of these duties, she could sunbathe, swim,

THE ELUSIVE BABOON

play tennis, or do nothing at all if she chose. In contrast, her English counterpart did her own shopping, gardening, housework, and cooking, while tending to the children. A baby was time-consuming and generated a lot of work, but if all tasks were taken care of by others, children presented little hardship. No one seemed concerned about the serious problems of population explosion, and families on ex-pat salaries lived very comfortably indeed.

Soon after we arrived, Cliff started an animal collection. He devised traps from empty baked bean cans, wire, and rubber bands, baited them, and every evening put them in the hedges. An animal entered; the door would snap shut and be held shut by a rubber band. Each morning, he went out early to check on them. Delighted by the extra space in the house, he set up some cages for rodents in our spare bedroom. They were not animals I liked, and I hovered on the edge to watch. But Caroline was intrigued and not at all afraid. We ended up with creatures of all shapes and sizes with interesting adaptations to their environment. One had two outer digits on its feet adapted for climbing while its tail was adapted for winding around twigs, allowing it to cling tightly. Two had a curious habit of doing backward somersaults and became a source of entertainment to our neighbors. Some of these small rodents had spots; others were striped. My favorite creature was a pygmy mouse about two inches long with prominent triangular ears. It liked peanuts and cornflakes. After a while it acquired a mate, but pygmy mice are prolific breeders. Soon we were inundated with them and returned most to the wild.

Two chameleons lived on a twig propped against the window. These "dear little things," according to Caroline, were bright green and looked like creatures from a science fiction movie. Large, independently flexible eyes protruded from their heads and had the unnerving habit of moving backward and forward as though constantly following you even when their backs were turned. Geckos with pale, transparent, fawn-colored skin lived in the roof but came out in the evenings and ran along the walls, clinging to them with pads on their feet. Tiny ants formed long straggling columns on the walls that looked like moving black cracks.

Cliff rigged up a mist net for catching bats in the garden. Every day he caught several and prepared them to send to Canada for a collection at the University of Toronto. He was paid a shilling for each. On our very tight budget, this helped enormously. Later, two monkeys arrived and lived on the verandah in cages, and eventually we added some bush-babies to the menagerie in the spare bedroom. Some neighbors thought we were mad; I could see why.

We continued to depend on others for transportation. Cliff wanted to buy a vehicle large enough to carry equipment and tough enough to negotiate rough ground in the field. After he returned from the States, desperate for a good cup of English tea, he found a Land Rover in Kampala within his budget and took it to a garage for an overhaul. On November 11, the day after Caroline's second birthday, I was preparing supper when I heard a vehicle halt at the approach to our driveway. For a minute there was silence. This was followed by the sound of gears crunching, the sudden roar of an engine, and from around the high hedge by the road, the front of the large gray Land Rover lurched into view. It wheezed up the slope of the drive like an old man with bronchitis before it came to rest in front of the garage. After our tiny blue minivan, it was gigantic. I dropped what I was doing and went to inspect and admire. The vehicle looked in fine shape except for a large area on the left side that had been patched up. I was told it had been charged at speed by an irate rhino. On contact, the rhino had ripped open the body with its horns before the driver managed to get away. True or not, I thought the patch lent our Land Rover an air of distinction and showed that it had had an eventful past. More than anything, it meant we could go on field trips and travel into town without relying on others. Now I had to learn the art of how and where to shop.

CHAPTER 12

Learning New Ways

RIKKI WALKER COULD BE OPINIONATED and was not averse to telling me what to do. Sometimes she told me stories that seemed far-fetched, like one about the human fetus she had kept in a jar on the mantelpiece. At times I resented her bossing me around, but I was grateful for her generosity, and comforted by the knowledge that she would always help out especially in medical emergencies. She was nine years my senior, had more lifetime experience, had been in Uganda longer, and knew more about how things worked. She said she would show me the best places to shop. "Buy a large rectangular basket with a looped handle, Jennifer. You'll need that for shopping in the outdoor market," she instructed me. I dutifully bought a basket.

On my first shopping trip, Rikki pulled up at our house and, with basket in hand, I climbed into her car and we set off along the Gayaza road to Kampala. First we stopped at the supermarket, where she said many ex-pats bought all their food, but we were not going to follow their example. We would buy commodities such as milk, biscuits, canned goods, cereals and peanut butter, but she had learned that meat and vegetables were much better from an Indian butcher and at the Kampala open-air market. I had to buy enough food for a week.

Our second stop was the butcher's shop. This was different from any shop I had ever come across in England, where the windows displayed meat, pork pies, steak pies, bacon, and sausages. There was no sign of meat in the butcher's shop in Kampala. You went through the door into an open space, approached a large wooden counter in the center, and an Indian in slacks, shirt, and a navy-blue butcher's apron appeared from somewhere in the back. You told him what you wanted before he opened the door to an enormous refrigerator behind the counter. As he stepped inside, you caught a glimpse of animal carcasses hanging from hooks in the ceiling before he closed the door behind him. He would reappear with a large chunk of meat, slap it on the counter, proceed to cut off the amount you requested, wrap it in paper, and immediately return the rest to the refrigerator before coming back for payment. His meat came in by train from the Kenya Highlands and was excellent. It cost so little that I spent only about seven shillings for a week's supply.

"Now," said Rikki, "we'll poke around a small all-purpose shop. Lots of stuff for curry, you know."

We got in the car and stopped by a small Indian shop or *duka,* the outside of which was quite unprepossessing—just a tiny window to the left of a small door covered with an awning. We got out, went to the door, carefully pushed it open, and stepped into a room where the interior could have been no more than about six feet square. Smells of spices and soaps greeted us in the shady, sultry interior. On the far left stood a wooden counter behind which two Indian men with well-oiled black hair waited attentively to serve customers. The counter rested on top of a glass-fronted display case crammed full of goods. There were Swan Vesta matches, aspirins and various other medications, bars of soap, packets of needles, reels of cotton, hair slides, hair grips, hairnets, shampoos, cigarettes, toothpaste and toothbrushes, glass beakers, teapots, cups

THE ELUSIVE BABOON

and saucers in heavy glazed pottery with a flowered design, and aluminum mugs, knives, forks, and spoons. Some had been there for a while, judging by the layer of dust. Bright yellow coriander and orange turmeric sat in canisters on the counter. A row of barrels filled with coarse salt, ground nuts, split peas, orange beans, and lentils for use in soups and a variety of Indian dishes stood on the floor in front of the shop window. Onions hung in bunches from the walls and ceiling.

Shelves lined the walls and were packed from floor to ceiling with fabrics, household goods, and medicines. Some contained canned goods, crockery, and cleaning materials, others rolls of plain cotton fabric in bright reds, oranges, blues, greens, and yellows, but some had flowers, stripes, or checkered patterns on a white background. There was constant activity as people drifted in and out to chat to the shopkeepers or purchase some aromatic substance from one of the many jars before departing with small packets wrapped in newspaper. We went to buy rice but, enticed by the smells, often ended up with a bunch of small packages of spices.

Our last stop was the open-air fruit and vegetable market. That was where I needed my large basket. Rikki told me that someone had to initiate you into the process of shopping there, and I could see why. Our first challenge was to park the car, because we were besieged on the outskirts by a crowd of teenage boys waving wildly and trying to direct us to a place in the already overcrowded parking lot. Once parked, they surrounded us, pushing and shoving, and so reminded me of the scene we had encountered at the Kampala station. Without any explanation and guidance about what to do, I would have been terrified.

I learned that you had to decide which boy would carry your basket. As you sat in the car, you had to point at one of those in the crowd and say, "You." He then leaped forward, shoved the others away, and the pack went off on the hunt for another car. Only then was it safe to emerge. For a shilling, a boy accompanied you in the market while you purchased produce. He transported it to the car, sometimes staggering under the weight of the basket. There was a great deal of competition for this privilege.

As we entered the market, we were greeted by a hive of activity. People were coming and going and shouting to one another. Stalls on the edge offered household goods and bolts of bright-colored cloth, but we headed for those in the center heaped with fruit and vegetables. Huge bunches of green *matoke* lay piled up on the floor. Everything had just come in from the countryside, was extremely fresh, of excellent quality, and, by our standards, very cheap. Nevertheless, we still haggled over each item. I could buy nine avocado pears for a shilling, about a thirtieth of the price in London. Delicious large mangoes were three for a shilling, and two pounds of tomatoes cost the same. Although we managed to reduce the prices, we were told *mazungu* (Europeans) always paid more than Indians, while they in turn paid more than Africans.

You could pick and choose, and we spent a lot of time squeezing, poking, sniffing, and putting things back before we made our final selections. I got so much into the habit of this that I landed in trouble when we returned to London and did the same thing in the Camden Town street market. There the stall holder stared at me in angry disbelief when I started to pick over his goods. His face grew redder and redder, as if he were about to have an apoplectic fit, until he shouted in a strong cockney accent, "Git yer 'ands off them tomatoes, missus, and stop poking at the fruit and spoiling the display."

I immediately reverted to being very British and abjectly apologized. But the stall keepers in Kampala were only too eager to sell and to please. They would press things on you, saying, "This is very good, madam," and if you rejected it, they would hastily find an alternative. We were even encouraged to taste things but avoided that. There was no indication anything had been washed, and you had to be careful about germs.

So while some of our *muzungu* neighbors bought their fruit and vegetables in the supermarket, the quality was far superior in the open air market, and I was very glad when Rikki introduced me to it.

THE ELUSIVE BABOON

In another part of the market, women dressed in long cotton dresses and bright head-kerchiefs in distinctive patterns of red, black, and yellow sat on the ground, chatting to one another, often while nursing a baby at their breasts. Each woman had a small pile of fruit, vegetables, and peanuts in front of her, and when we went past she would reach out to touch us to try to sell some of her wares. In the midst of this, men in grayish-white *kanzus* arrived from the countryside on bicycles with bunches of *matoke*, bananas, or cages with chickens strapped to the back. I was fascinated to see a busy trade in live chickens taking place.

We once ventured across the road from the fruit and vegetable market to look into an open-air meat market used exclusively by the local people. Stringy meat from emaciated cow carcasses lay on the ground, and men were hacking it to pieces. Numerous vultures with their bare heads, long necks jutting forwards, and beady eyes flicking back and forth waited in the trees for their opportunity to dive down on the rotting entrails after people left. The meat may have been freshly cut but was a stomach-turner for me, and I didn't linger.

Toward mid-day we finished haggling and poking. The sun was getting high in the sky and we headed back to the car, paid the boys with the baskets, and flopped down on the car seats, which were now hot from the sun's direct rays. There was only one hitch in the whole procedure: Rikki insisted on going to the butcher before the outdoor market, and we would leave the meat in the car. I was afraid it would go bad in the heat. She assured me we hadn't left it for very long, and that beef, which was mostly what we bought, was perfectly edible even if it had gone off. As long as meat was thoroughly cooked, the bacteria would be killed, so why worry? I argued that, even if she was right, I didn't want to eat meat that might taste tainted. She countered, "We are surrounded by Indian shops packed with spices, and you can overcome the taste with curry."

I said I wasn't sure I would fancy that because I would know the spice was covering up tainted meat. She laughed and said, "I know you don't believe me,

Jennifer, but the meat would be fine."

"Whether I believe you or not, I *still* don't like the idea of leaving meat in a hot car."

In the end I always lost the argument, because we went in her car and continued with our routine. She wanted to shop early and thought it better to visit the open-air market after completing our other shopping because there was a bigger selection of fresh produce later in the morning. After a while, I relaxed. As soon as we got back I put the meat in the fridge, and it was so fresh when we bought it, there was never a problem and it always tasted good. As a result, while the rent took most of our money, good food cost very little and we ate well.

So I learned the art of food shopping. It was time to find out more about the Africans who lived on our estate and in the surrounding areas. This proved to be an eye-opening experience.

CHAPTER 13

Local Characters

AFTER JOHN AND FRED SETTLED into their own quarters, we heard the clang of pots and shrieks of laughter floating across the air as they cooked and entertained friends. In the mornings they often stopped work to talk to the many Africans who came through our garden, which it turned out was a thoroughfare going from a village somewhere at the back of the house to one in the valley beyond. Long drawn-out greetings of *Jambo, Habari* and *Mzuri,* followed by "ahs" and "ayes," filled in silences before they burst into further conversation. You raised your right hand when greeting people. To use your left was an insult. This was because the right hand was used for the cleaner things in life such as eating; the left was used for cruder activities, most notably wiping one's rear end "after you've been to the crapper," as people said.

Fred kept the grass under control, stopping every so often to pass the time of day with people who tramped across the garden. We had no lawn mower, but he was adept at cutting the grass with a lethal-looking slasher consisting of a wooden handle and a flattened iron blade about two inches wide and two feet long with a slight curve at the end. He swung it backward and forward like a pendulum, inscribing an arc from a point slightly above the shoulder on one side of his body to an equivalent point on the other. To cut more closely, he used a flicking motion of his wrist. I watched him working and thought the repetitive activity must be boring, but he didn't seem to mind. He could break the monotony with his conversations, and by doing a bit each day he kept the huge stretch of grass tidy.

Fred dressed in clean khaki shorts, a short-sleeved white shirt, and rubber sandals with thongs. He had an open face and a smile that was slow to come, but his face lit up and he revealed beautiful, even white teeth when he was amused. He came to us from a village several miles away. His schooling was limited, and he spoke and understood very little English, but with the help of others we communicated effectively. I discovered he was a quick learner who soon began to understand quite well. He listened with his eyes downcast, frowning slightly as he concentrated. A big smile spread across his face when he thought he understood, and, based on his actions afterwards he was generally right. He was quiet, self-contained, charming, polite, and trustworthy.

John, our houseboy, knew more English but was a scatterbrain who paid much less attention. When I asked him to do something, he rushed off before I finished talking. Often he caught one word in a sentence and filled in the rest for himself, which led to some odd behavior and numerous misunderstandings. One of his most annoying activities occurred soon after we arrived, when I asked him to wash the floor in the bathroom, which was covered with red dirt brought in by visiting children. "Yes, madam," he said and immediately took off. Suddenly a flood of water shot from under the bathroom door and John came out with bare feet. He had tipped a bucket of water onto the floor and rubbed it around with a cloth, but instead of mopping it up and squeezing the water into the bucket, he had removed the deluge by putting his cloth in the water, walking backward, pulling the cloth along, trailing water across the floor and leaving a wet strip behind him. In this manner he went down the passage by the side of the bedroom, past the dining area, through the kitchen, and out the kitchen door. Then he kept going back to repeat the process. It seemed to me he did little more than spread the muck around, but no matter what I said, he continued to employ this messy and inefficient method to clean floors. We had constant trails of water all over the house.

Sometimes I asked him to complete a task and he did something completely different. "John, the table is very dirty. Please, would you clean it for me?"

"Yes, madam."

I would wait but nothing happened, or I might find he had put the kettle on or done some other totally unrelated task.

I once attempted to speak to him in Swahili. This sent him into fits of laughter, and he claimed to have no idea what I was talking about. I had been trying to learn Swahili from *Teach Yourself Swahili* and the small book called *Up Country Swahili* that Cliff had unearthed in London. My copy had been printed in 1964, thirteen editions after it first appeared in 1936, and it was not up-to-date. Some parts were relevant and useful; others contained odd phrases such as "The cook is beating the witch doctor" and "I am hitting the drunkard with a stick." Eventually I gave up trying to speak to John in Swahili, and we soldiered on in English.

John's floor cleaning was a constant aggravation, but some of his other behavior positively horrified me. He once used a whole packet of washing detergent to clean a small vegetable pan. We normally used such an amount in a month to wash our clothes. Then he decided to clean the floor with Ajax scouring powder. He sprinkled it liberally all over the place, failed to wipe it off properly, and we all ended up with sore feet.

Soon after we acquired the Land Rover, John said he had bought a bed and asked if we would drive him into Kampala to bring it back on the Land Rover. A bed was a prized possession made with a wooden frame and given its springiness from rubber laths made from old car inner tubes that crisscrossed. On the appointed day, John, Caroline, Cliff, and I piled into the car together with Fred, who came along to help. John sat in front to give directions and waved like royalty to people we passed on the road.

Seven miles later we reached Kampala, where it became abundantly clear that John had no idea where to find the bed. He climbed out, talked to people, inquired around in a haphazard fashion for some time, and then came back with a tale about the bed not being ready until "Thursday." We wondered whether the bed even existed or this had been a good excuse for an outing, and all he wanted to do was go for a ride and wave to his friends. Fred was concerned that we had gone all that way for nothing. John could have cared less. Some weeks later he asked once more to be taken to collect the bed. We told him this was his last chance. If there was no bed, we were not going again. Our threat seemed to sink in, and he postponed the trip for a few days before we set off for town. This time the bed materialized, and we transported it back on top of the Land Rover.

Eventually John had to be released from his duties because he was unreliable and would take our things without permission. We heard rumors that he had been caught stealing and been dismissed from previous jobs, but he'd come with good references and was prepared to work part-time, which suited us. Later, we heard tales about him that were hard to believe. One of the houseboys claimed John was the leader of a gang of hoodlums who went around stealing and terrorizing the neighborhood. John was small and thin and probably about fifteen years old. He generally had a big smile on his face. It was difficult to imagine that, beneath his smiling exterior, lurked a treacherous thug and gang leader. Besides, the man who originated the story was from Kenya, and we heard that the Kenyan help and the locals disliked one another, so we questioned the source.

Fred took over John's duties, which were not extensive. He was pleased to have the extra money. John thought he could come back when he felt like it and started to hurl abuse at me when I refused to let him. I began to wonder whether there *was* some truth to the rumors about him and was relieved when he departed for good.

We also came to know Rikki's houseboy, Lawrence, a stocky young man about five feet six inches tall, with broad facial features. I thought he saw himself as a grand interpreter and organizer. Lawrence was very pleasant but serious and intense. He loved to gossip about himself and others. This led to some rather odd conversations such as the one when he told me his gonorrhea had got the better of him and he needed to take time off to get some medicine.

I wondered whether I had heard him correctly. "What did you say is a problem, Lawrence?"

"My gonorrhea, madam."

"I see. What does Mrs. Walker say about this?"

"She was *most* interested, madam."

"Really!" Surprised by a revelation of so personal a nature, I asked no more questions but could imagine Rikki getting the lowdown before she gave him time off with money for medicine.

Lawrence also told me that Fred, like many of the boys, was trying to save money to go to school. Fred appeared to have good intentions until the temptation of a transistor radio proved too great. As usual, Lawrence was in on the act, saying, "One of these men sell it for very little money, madam."

The next week the radio broke down, causing a great outburst among the locals. Everyone sympathetic to Fred declared the radio had been sold by a crook. The previous week they had claimed the radio came from a reliable source. A huge feud sprang up, with people hurling abuse at one another. We quickly distanced ourselves.

Lawrence added fuel to the fire about John with stories like the following when he came to the house after we arrived back from safari. We had exchanged the usual greetings before he said, "Madam, I think you should know John was down at five mile wearing Dr. Jolly's boots."

I looked at him in disbelief, because the said Dr. Jolly was about twice John's size. The boots would have been far too large. And Cliff had taken his boots with him. I was sure the boots were mine, the ones with the fleecy linings that I hated. John must have been a comical sight like Mickey Mouse with the large boots on the end of his skinny legs.

"Lawrence, are you quite sure about this?"

"Oh, yes, madam, John showed these boots to other people. John is a very bad boy, madam," he declared self-righteously.

An elderly man regularly strolled across the garden. Caroline called him "Jumble." This was her version of "Jambo," which he always shouted in greeting from the distance, but it could well have described his appearance. Jumble was quite tall and thin. He dressed in an incredibly tatty off-white *kanzu* which came to his ankles, but his headgear set him apart. This consisted of an old bottle-green rubber bathing cap like the ones women wore to protect their hair when swimming during the 1940s. I called it his Esther Williams hat. He wore it inside-out and perched on the back of his head over his grizzled gray hair. The rubber chin-strap with its metal clasp dangled next to his small gray beard. Jumble shuffled along in his long robe with a canvas collecting-bag slung across his left shoulder. In his right hand he carried a stick that he used to poke around the dustbins. He must have unearthed the bathing cap when he rummaged through the bins after someone threw it out, and he was obviously attached to the odd acquisition, for we never saw him without it.

Like most Africans in the vicinity, Jumble was a great collector of empty bottles, tin cans, and brown paper bags, because they could be sold or used for storage. Bottles were mostly used as containers for cooking oil, paraffin, or *waragi*, the potent local banana brew. Tin cans were highly prized for use as cups or for storing items such as beans or sugar. I once went out to try to talk to Jumble, but when he saw me advancing he headed off at speed, surrounded by a flurry of tatty robes, and always kept his distance.

As a clock-watcher, I found the Africans' slower pace of life and lack of attention to time a challenge. They seemed to sit around talking for hours, never appeared to hurry or rush, and often didn't turn up at an appointed time. But Ugandans thought about time, place, and people in a very different way from us. A man once told me he would bring me some tomatoes "tomorrow at four, madam." Five days later, by which time I had all that I needed, he appeared, beaming. "Here are the tomatoes I promised, madam." He held out his hand for payment.

It was like asking the way. Africans would say "over there" and point vaguely in the air. If the destination happened to be a long way off, they would just point higher, which conveyed very little to us.

House boys and *shamba* boys were constantly in financial trouble. They were tempted to buy things they could not afford and endlessly tried to borrow money from their employers or other house boys. Wages were accounted for months in advance, but arguments about who owed what to whom never ended. At times things became complicated, as when Lawrence put himself down for the Nile School and borrowed money from Rikki to help pay for it. Meanwhile John had borrowed money from Lawrence and from us to purchase a three-speed bicycle. Lawrence now demanded that John pay back the money he had loaned to him so he could put it towards his schooling. But John had no money because he had spent it all when he made a down-payment on the three-speed bike, which then didn't work. A great feud sprang up between Lawrence and John, with Lawrence reminding me yet again that "John is a very bad boy, madam."

As in many other societies, kinship and other relationships were expressed in categories unlike those in the familiar European system. Mother's sisters and father's brothers were referred to as "mother" and "father." Boys who came to chat often referred to one another as "brother" or "cousin" when they were not blood relatives. If they had the same father and different mothers, they said they were brothers, not half-brothers or step-brothers as we would have done. Fred, however, had a sister who visited from his village. She looked exactly like him—"Same mother, same father, madam."

One day after John and Fred had finished work and Caroline was with friends, a man came to the kitchen door and claimed he repaired canvas and leather goods. Small and wiry in build, he peered at me intently through round wire-rimmed glasses. He was clad in a white shirt, long baggy gray trousers, and old leather shoes with holes. His head was completely bald. He demanded I give him some work, yet I had nothing in need of repair. But he refused to go away and stared so intently I felt threatened. It didn't occur to me that he was very short-sighted. No one was around, and I was so unnerved by his look that I told him to wait while I went into the house, where I found an old leather hold-all that was losing its zipper. I headed back. He said nothing but snatched the bag and before he threw it to the ground, peered at it so closely that his nose almost touched the leather. Then, to my horror, he started to take off his clothes.

I stood rooted to the spot while he proceeded to remove his battered shoes to reveal a pair of very dirty white socks full of holes, which he took off to expose a pair of extraordinarily grimy feet. Next he took off his shirt and then—I could hardly believe it—he started to remove his trousers. At that point, he was standing in a pair of strange grayish-white cotton shorts which, to my intense relief, he kept on. They reminded me of the shorts worn by the All-England Ladies Hockey Team in the early 1950s, because they had pleats in the front and flapped around the knees. Satisfied that he was ready, the man sat cross-legged on the ground just outside the door and began to mend the bag. He used a large needle and khaki thread, and all the while held everything only a few inches from his face. After some intense concentration he finished and had done a reasonable job though it made little difference to the utility of the old bag. Needless to say, he demanded an exorbitant amount in payment, but as usual, we bargained and reached a compromise. I was very glad when he got dressed and I saw the back of him.

People wore interesting and varied clothing that ranged from traditional to mostly European. Women who could afford to do so mostly dressed in a style originally introduced by missionaries. This consisted of a long dress in patterned cotton of bright colors such as reds, blues, greens, and white, with a large fold or loop at the waist. The top generally had a square-cut neck and short, puffy sleeves. Many women wore colored headscarves tied in a knot at the back of their heads. Generally, they wore nothing on their feet. Small babies were carried on the mother's back in a strip of cloth which tied around the woman's waist and over her breasts. This freed her hands for work or for carrying other goods.

Many of the men wore interesting hats, including old trilbies that were discolored and battered, with felt crowns and brims that had shrunk. You would see hats like that on scarecrows in England, but any man wearing a trilby seemed very pleased with his hat and walked with a jaunty air. Some hats were very small and perched on the top of heads. Others were too big and sat on their wearer's ears. I once saw a man riding a bicycle who was incongruously dressed in a flowing white *kanzu* and a hat like one that an English woman would wear at a garden party. It was made of fine straw, had a wide brim, and was decorated around the crown with a sky-blue ribbon tied in a bow at the back. Some hats were shaped like lampshades and others like pork pies. All were quite distinctive. Shoes, as I have said, were uncommon, although people occasionally wore shoes made of plastic or old car tires, which must have been uncomfortable in the hot climate. Some men wore *kanzus* but others wore shorts and shirts. I saw some very poor men in shirts that were so torn and tatty they hung in ribbons and I wondered how they managed to get into them. I even saw one in a backless shirt held on by a strip of cloth around his neck.

Generally, if a man had money, he wore a smart shirt, long trousers, and shoes like the "Nelsadry man." Nelsadry was the name of a supermarket on Kololo Hill in Kampala, where he worked. The Nelsadry man arrived once a week to deliver a treat—an airmail copy of *The Observer* from London. He would roar up the drive in his white van, screech to a halt, leap out, and swing around the corner, well-dressed in his long gray trousers, white shirt, and leather shoes with laces. After beating a *rum-ti-tum* on the door, shouting a cheerful, "Jambo," and delivering the paper, he had a quick word with the boys and jumped back in his van. The engine roared to life, and tires screeched as he shot down the drive before he turned onto the road and disappeared for another week.

 Together with the two cheerful *askaris*, or security guards, who stood near the entrance to the estate and wore red fezzes, khaki shirts and shorts, and black knee-socks, these were some of the Africans we saw on a regular basis. And then we got to know Jimmy Kitumba.

CHAPTER 14

Jimmy Kitumba

ALTHOUGH CLIFF'S MAIN AIM was to collect baboon blood, he also drew blood from vervet monkeys for analysis. Many of his samples came from a monkey trapper and exporter at Entebbe. But he decided that, rather than depend solely on others, he would trap some vervets himself. For this he needed help. When word got out that he was looking for an assistant, Lawrence was first on the scene bringing with him one of his many relations. His name was Jimmy Kitumba. Jimmy was about sixteen and, like everyone else, wanted to earn money for school by working various jobs. He was keen to become a businessman. As a monkey-catcher his qualifications were twofold. First, he could speak good English. Second, he claimed vervet monkeys surrounded the village in which he lived, and he assured us it was possible to catch them. We hired him.

Although said to be related, Jimmy and Lawrence were a study in contrasts. Lawrence was stocky, with rather broad features and a snub nose. Jimmy was thin, with smooth brown skin like that of a shiny, newly emerged horse-chestnut. His nose reminded me of John Lennon's, being narrow at the top with nostrils that flared out at the bottom. Lawrence loved to gossip about people and generally put a negative spin on his stories for effect; Jimmy was curious and wanted to find out about people and places.

Once hired, he turned up for work neatly dressed in a white cotton shirt and khaki trousers belted at the waist. He liked to sit down with me and ask questions while waiting for Cliff to appear. I would see his brown eyes peering through the French doors until he spotted me. Then he knocked, entered, sat on a chair in the living room, and began to talk. As he did so, he mused over his problems and shared his dreams. He liked to speculate on life in London and confided to me that his ambition was to go there. Not surprisingly, he was very naïve about city life in other countries and especially in a city as large as London. The largest place he knew was Kampala, which was extremely small in comparison. He would ask me about London and how we lived. When he did so, Jimmy had an odd habit of making a statement in which a question was implied but not asked directly. "Madam," he said, "I think you have a big garden and need a *shamba* boy when you are in London?"

"No, Jimmy, I have a very small garden there, and I take care of it myself. Only rich people can afford to have someone take care of the garden for them in London."

This seemed to go over his head, for he persisted: "But I should like to know what your *shamba* boy is like, madam."

"I don't *have* a *shamba* boy, Jimmy. Wealthy people like the Queen have someone to take care of their gardens. I don't. I work in my own garden."

He frowned and obviously didn't believe me because he continued to make references to my non-existent garden boy. Then he switched gears. "Madam, I think your houseboy likes England?"

I had no idea how to answer this. Was he trying to find out what it was like to be a houseboy in England? Was he trying to find out whether he could come to England with us when we returned? Maybe he was saying, *I think I would like England if I were your houseboy.*

The only answer I could think of without getting into complications was, "Jimmy, I don't have a houseboy in England. I do the cleaning, washing, and shopping by myself."

"Madam, I think that is impossible! I think you must have a houseboy."

I gave up because I realized Jimmy had rarely seen European women do things for themselves. In Uganda they all had houseboys, garden boys, and *ayahs*. Some did no work apart from shopping. He sat and mused for a while before he said, "Madam, I must save enough money to go to London. I wish to find work there."

To me, this hardly bore thinking about. I imagined what might happen if we took him back to London with us. Maybe he thought we had plenty of room with a house like the one we were occupying in Katalemwa, and he could stay with us. He would have been shocked to see our small two-bedroom duplex in Camden Town. He had already indicated he thought we had a garden boy and a houseboy; he would have had to absorb the reality of these non-existent people. He would have needed different clothes. It was almost impossible to imagine taking him into an office and introducing him: "Hello, this is Jimmy Kitumba."

"Oh, hello, Jimmy."

Then he would look wide-eyed and start to ask his strangely worded questions. I was afraid people might see him as a novelty for a start but would soon get impatient and ignore him.

Maybe he could realize his fantasies about being a businessman in Uganda, where he spoke the language fluently, but in London I could think of nothing he could do apart from menial tasks such as cleaning floors or washing dishes. He would be shocked by the cold weather and would have no friends to chat with on a daily basis and pass the time. I felt he would be sadly disillusioned by what he found and would soon feel very homesick for his own country with its greenery, conviviality, lack of traffic, slow pace, and absence of concern over the clock.

He would have no idea how to negotiate public transport. Jimmy's idea of rapid transit was to go for a ride in our Land Rover. How could he understand coal fires, frost on the windows, or snow on the ground? I felt sad to see someone so smart and eager to get ahead with no means to do so, but at the same time I envied him. He lived in beautiful surroundings with friends and family all around who welcomed him. People shared what they had and took care of one another. In many ways he had a good life and he was used to it, so I said, "Jimmy, I think that you have a beautiful country. I think you would not like London. People are not so friendly. You would miss your own friends, and I think you might be very sad."

"I think that is not so, madam."

With the language barriers between us, how could I go into details about the problems he would encounter? In any case, I felt sure that he would never have believed me about the damp, the bias against black people, problems in finding accommodation, the Underground railway, and all the other unique features of a big city incomprehensible to a person who has never experienced anything of the kind.

He once asked me about the washing habits of people in England: "Madam, I think it is true that the price of bath water is very high in England. I have been told that people only wash the face and hands, that they do not often wash the body."

I was tempted to laugh but stopped when I saw his serious face and tried to think what to tell him. There was some truth in what he said, but the expense of bath water was only part of it. His question triggered a lot of memories. Many people in my generation had grown up at a time when there were no fancy bathrooms and only limited amounts of hot water for bathing. Water was heated by immersion heaters powered by gas or electricity. In my parents' home we had electric and gas meters under the stairs that had to be fed with shillings to keep the lights on, gas in the cooker, and the immersion heaters running. I could have told Jimmy that putting money in the meters could get expensive, and that this made bathing water costly, but I wasn't sure how much he would understand.

We'd had no such things as showers when I was small and were limited to one bath a week, washed our hair once a week, and sponged ourselves off in-between. We must have smelled awful. Maybe Jimmy had heard a story like that, and this was what he was talking about. However, he would never have understood the austerity measures in which we grew up after World War II, and how we'd had to economize on clothes, food, and heat for water. Anyway, bathing was darned cold and uncomfortable in most English bathrooms, especially in winter without central heating. How could I explain all of this? I was wondering when I suddenly realized he was silent and waiting for me to answer. I gave a very inadequate response. "Well, Jimmy, it can be expensive to heat water in England, but it's also cold for much of the time. Most people don't want to undress in the cold, so they just clean the bits of skin that show outside their clothing."

Jimmy's only comment was, "Oh, I see." I knew he didn't understand any of it.

Jimmy loved to ride in the Land Rover, where he sat up front on the passenger seat during monkey-trapping expeditions. He and Cliff would head to an area near to Jimmy's village where several traps had been set to catch the vervets. These traps often caught other creatures, and Jimmy would see them when he checked the traps. On one occasion he remarked to Cliff in a thoughtful tone, "We caught one of those pine-paws in your trap, sir."

"What did you say?"

"Pine-paws, I think it is."

Cliff thought for a minute before the penny dropped. "Pine-paws? I think you mean porcupines."

An enormous smile lit up Jimmy's face. "Yes, that's it. We caught one of the pork pies."

Cliff replied, slowly and deliberately, "Por-cu-pines."

"Yes, sir, that is what I am saying to you, pork pies."

Rather than spoil Jimmy's great satisfaction and pleasure at having labeled the strange creature, Cliff decided to let the matter rest and hid his amusement by remarking, "That's terrific. Let's go and find it, shall we?"

Of course Jimmy had never seen a pork pie. No doubt he would have been amazed to find it was a roundish object with a baked crust wrapped around cooked ground pork, a favorite food with many English people.

A big highlight in Jimmy's life occurred when he went to Kampala to see the Beatles in *A Hard Day's Night*. He was mystified by their long hair and accents. Most white men Jimmy met had short back and sides, even crew cuts, while he and his friends had close-cropped curly black hair. When I thought of the mop tops of John, Paul, Ringo, and George, and of their Liverpool accents, I could see why he was puzzled. Nevertheless, Jimmy thoroughly enjoyed the music, and the movie was a great treat for him. He beamed when he saw me and said, "Madam, this was a very fine film with these strange Beatle men!"

Eventually we became very fond of Jimmy. He was generous and kind and a great help with the vervet work. However, our main mission was to trap baboons—and this was proving to be a major challenge.

CHAPTER 15

Elephants Galore

WHEN WE SAT AT OUR DINING TABLE in London and said, "We'll trap baboons in the wild," Uganda seemed an ideal place. Never could we have anticipated the difficulties we would encounter.

A major reason for choosing Uganda was the variety of its terrain, with baboons located in different habitats. It is a land-locked country covering 91,134 square miles and about the size of Ireland. Bounded by the Sudan, Kenya, Tanzania, Rwanda, and the Congo, it forms a dissected plateau of from three to five thousand feet, rimmed by mountains, old volcanic peaks, and lakes. Part of the Great African Rift valley, it contains Lakes Albert, Edward, and George along its western border, with the Rwenzori Mountains, or "Mountains of the Moon," south of Lake Albert and north of Lake Edward. Further south, the extinct Virunga volcanoes, home to the mountain gorilla, border Rwanda and the Congo. The equator cuts across Lake Victoria just south of Entebbe, and the mountains of the Kenya border rise in the east.

Extensive areas are covered in savannah grasslands where you see the big game, most notably in Murchison Falls National Park. Other areas are covered with rain forests, like Budongo, that harbor their own distinctive plants and animals. In the center of the country, Lake Kyoga, large and shallow, stretches its finger-like tentacles eastward and is surrounded by extensive areas of low-lying, mosquito-ridden swamps. In the north it becomes much drier, with Karamoja, a dry grassy area, in the north-eastern corner. Throughout the year the temperature varies very little and is generally in the high seventies or low eighties, though it can go much higher. There are two rainy seasons: one lasts from March to May, the other approximately from September to November.

We had chosen three diverse areas for trapping. One was the West Nile region, where baboons lived in hilly areas covered with scrub, trees, and brush. Another was the Budongo Forest in the west and about three to four hours away by road from Kampala; the third was in the southwest, where the animals were found in savannah grasslands and rolling hills.

Toward the end of 1965 we decided to take our first trip to the West Nile region, about four hundred miles away. It was December and, being the dry season, one of the best times to travel. But first we had to get permission from the local game warden and a police permit to visit because of recent border incidents, with the Sudan and Congo close to where we were headed. We checked at the British High Commission to see whether the area was safe. They anticipated no problems.

By then Cliff felt confident that he knew how to deal with the baboons. He had practiced on some that were kept in a large cage on Makerere Hill by Thelma Rowell, a visiting Cambridge zoologist who was studying their behavior. With a team of people to help, her baboons were lured to the edge of the cage with bait and kept there long enough for Cliff to sedate them, with a drug injected from a syringe at the end of a pole, and draw blood. At that time, Sernylan was the drug of choice because it rendered the animals immobile within minutes and had no long-term effects. Short-term they had to be watched to see they came to no harm, because when the drug began to wear off they rolled around like drunks and their movements took place in slow motion.

All went well until the caged baboons who were not sedated got fed up with what was going on. They started to threaten the humans through the wire of the cage by biting at them, scratching them, or pulling their hair if they got too close. But Cliff got his blood samples.

Now he was going to put what he learned with the caged animals into practice with wild ones.

We were going for five days, would have no refrigeration, needed to take all our food with us, and so stocked up on canned foods and dried goods. Our canned foods were mostly corned beef and sardines; the dried goods were milk, sugar, tea, coffee, and various cereals. We took oats for porridge, and this eventually became a breakfast staple, while rice and spaghetti became staples for the main meals. We took some fresh fruits and vegetables, but when they ran out we decided to manage without. To buy them at the roadside was risky because everything, apart from bananas, needed to be washed carefully, and the water itself could be unsafe. We took a large canister of water with us from Kampala, because water was hazardous to drink in the field. When that ran out, we boiled water or used sterilizing tablets.

Before we arrived, I had unrealistically assumed someone would take care of Caroline when we went on field trips, but now I had no intention of leaving her. Either I would stay with her at Katalemwa, or she would come with us. My main fear was that, if she came to the field, she might get sick, and we would be some distance from good medical facilities. I never thought about any danger from my pregnancy. It just seemed a pity to lose the chance to see such a wonderful country. And so we all went while I was still able to travel.

This decision proved well worth it. Caroline turned out to be a great asset because she was friendly and a novelty to the local people, making it easier for us to get to know them. She also adapted extremely well to camping and the odd cooking arrangements.

On December 18, we checked once more with the British High Commission and, as they still said it was safe to go, packed the Land Rover with our provisions and camping equipment, had an early lunch, and headed in a northwesterly direction out of Kampala towards Masindi, about 130 miles away.

For a few miles we traveled on tarmac, but this soon petered out to become a *murrum* road composed of a stony surface and red laterite soil. This slowed progress, but we drove along without incident, passing women carrying bundles of *matoke*, bananas, wood, and other goods on their heads as they walked on the side of the dusty road, while men rode on bicycles with goods strapped to the back. By early evening we reached Masindi but, rather than setting up camp, treated ourselves to a night at the Masindi Hotel. Built in 1923 as a transport point for goods and produce from North Congo and Southern Sudan that were destined for export to European markets, it was Uganda's oldest hotel and full of history. Famous people had stayed there, such as Hemingway while on assignment, and Bogart and Hepburn while filming *The African Queen*.

"See," I said to Cliff, "my idea of wearing head-gear like Katherine Hepburn wasn't so outrageous after all."

He laughed, "This is not like being in a film, Jen."

"No, but I think there's something to be said for the wide-brimmed hat tied under the chin with a fine silk scarf."

"I'm not sure about the scarf, but you're right, we probably should have something for our heads."

That evening we cooked dinner in our room to save on expenses, but the next morning ate a hotel breakfast of fruit, bacon, sausage, and eggs, plus toast and marmalade, served with hot tea. It was delicious, and we filled up because it would probably be our last good meal for a while.

"I tell you, Cliff, I could get used to this very quickly."

He was munching on his eggs, bacon, and sausage with obvious relish. I thought he seemed to weaken a little and think about a more comfortable life.

THE ELUSIVE BABOON

"I know. There's nothing like a good old English breakfast."

"And how about a comfortable bed rather than sleeping on the floor of a tent?"

"That too, but you can't do my kind of work from a hotel, unfortunately."

It was true: His research could not be done in comfortable surroundings. The animals were in more remote parts, and while Cliff enjoyed the break in the hotel, comfort wasn't essential for him. Devoted to his work, he always showed dogged persistence and fortitude once he made up his mind to do something, and pressed forward even when conditions were bad. He reminded me of the early African explorers who wanted to discover new territory or convert the "heathens" to Christianity. Like them, he kept going even when illness struck, food was almost unpalatable, and the terrain was rough. Some years later, I found out that others commented on Cliff's resilience under adverse conditions. While they complained and moaned about bites from fleas, bedbugs, and mosquitos, Cliff seemed to tolerate them. Doug Cramer, an old friend who had served in the navy and was a champion weightlifter, once said after a field trip to Ethiopia, "I was itching like hell from these damn bedbugs that were eating us alive, but Cliffy said to ignore them. I was sharing a room with him, and he just rolled over and went to sleep. That S.O.B. can put up with any inconvenience."

I wondered if Cliff either had a high tolerance for pain, was able to exercise body and mind control like people who walked barefoot through burning embers, or whether both applied. For us lesser mortals, the conditions he tolerated were difficult. I preferred a softer and more pampered existence. The Masindi Hotel provided a pleasant interlude and was well worth the expense because we rested up and were ready to face the journey ahead.

Full of breakfast, we packed the Land Rover and were on the road before eight, heading towards the Murchison Falls National Park. We were to cut straight across it in a northerly direction. The weather was pleasantly cool as we drove towards it through a rather bushy region where tall grasses grew on the roadside. Gangs of workmen with scythes, who were slashing at the vegetation to keep it under control, stopped to watch as we went past. For the first time we saw the strange sausage trees whose fruit-pods hang down from long rope-like stalks and are shaped exactly like the long Italian or German sausages you see hanging in the windows of butchers' shops.

And then the area became much drier, and the trees thinned out, as we drove to the top of the escarpment overlooking the park. We stopped, got out, shaded our eyes, and looked down over vast areas of grassland. Ahead of us, the road wound precipitously down the escarpment. There were no other signs of life. The only sounds came from the clicking of insects.

We returned to the Rover and took the steep, rough road at a steady pace down to the gates of the park, where a cheerful guard greeted us. We paid him and entered. It was starting to get hot, but we were too excited by the wildlife to be much troubled by the heat. Hundreds of elephants wandered around and we had to stop when one enormous bulk crossed the road ahead of us. A sign read *Give Way To Elephants*, which amused us because there was little doubt about who should give way. The park was so full of elephants that we found ourselves close to several herds. Not until then did I realize how huge they would be at close range, and my excitement gave way to trepidation.

Unlike Cliff, who probably knew all the scientific facts about elephants by the time he was five or six, I had gleaned my information from children's picture books, zoos, and circuses. In my ignorance I hadn't differentiated between African and Indian elephants, and conjured up pictures of useful, friendly creatures. They let people ride on their backs and hauled logs. They ate buns offered to them at the zoo, and their looks provided a fascinating reminder of prehistoric ages. Decked out in colorful, ridiculous party hats and frilly clothes, they performed tricks at the circus when they sat on their hind legs and raised their trunks. Then there was that charming little French elephant *Babar* who, like a good citizen of the French Colonial Empire, dressed in Western attire and who "lived among men and learned much." I'd concluded elephants were accommodating creatures who did as they were told. But this was before the evening we spent with friends at Katalemwa, when the conversation turned to elephants.

After dinner, the husband settled back, crossed his legs, and poured himself a tot of whiskey before he said, "I expect you heard the story of the Volkswagen and the elephant?"

"No, what was that?"

He took a sip of whiskey. "Well, there was this chap driving through the Murchison Park in his Volkswagen Beetle. It's a tough little car that does well out here, you know. He's going along quite happily when he comes across this elephant with its back to him and its large behind blocking his way. He stops the car and waits for it to move, but the elephant doesn't budge. After about fifteen minutes, the driver gets fed-up and thinks, Maybe if I give it a nudge, it'll move and I'll continue on my merry way. He starts the engine and moves slowly forward, hoping the beast will hear him and move its great bulk so he can get past. The elephant stays put. So the driver goes up to the elephant and pushes it from behind. But the elephant's knees and the height of the bug's bonnet matched up. When he pulls forward and catches the elephant behind the knees, it does what any self-respecting elephant would do if pushed there. It sat down, landed on the bonnet of the little red bug, and squashed it."

I was wide eyed. "What happened next?"

"It seems the elephant righted itself, recovered its dignity, and lumbered off. That driver was damned lucky, if you ask me. He drove away and survived to tell the tale because the bug has its engine at the *back*, so it wasn't squashed. But that elephant could have tossed him and the bug into the air like the minibus the German couple drove."

". . . What was *that* about?"

He took another sip, leaned back, and continued, "These German tourists, a man and his wife, drove a minibus into the Queen Elizabeth Park where notices everywhere say *DO NOT Get Out Of Your Car*. That's because elephants are dangerous. Funny thing is that, in the midst of this, they have areas marked as camping grounds, yet elephants have obviously been there because there are lots of drunken-looking euphorbia trees surrounding them."

"What do you mean?"

"I mean they lean heavily to one side, and that's a sure sign that elephants have pushed against them. Anyway, these Germans saw the camping sign, thought it must be safe, and decided to set up camp. They parked, took out their tent, put it on the ground, and were about to erect it when this damned great elephant appeared out of the blue, saw their minibus, seized it, and threw it up in the air. Needless to say, the Germans were scared out of their wits. With the adrenalin pumping like crazy, they took to their heels and ran like hell. You wouldn't think it, but elephants can move fast. Lucky for them the elephant didn't run after them or they'd have been dead meat."

He stopped, reflected and took another sip. "It was also lucky Mweya Lodge was not far away. The tourists managed to reach it and rushed in like crazy folks. By then they were almost incoherent. They kept screeching out phrases in German, and nobody understood what had happened until later. After they were given a large dose of scotch, they began to calm down. Moral of the story: Never trust an elephant. They're unpredictable buggers with mean little eyes and bad tempers."

"Do you know what happened to the minivan?" I asked somewhat irrelevantly.

"Don't know any more about it. Maybe the elephant trampled on it."

He seemed to be enjoying himself, and I wasn't sure if he was pulling my leg. His stories seemed a bit far-fetched, but I knew people said elephants were mean and could run fast in spite of their bulk.

Now we were surrounded by these gigantic creatures, and I felt our Land Rover had shrunk to the size of a small matchbox toy. There were no other humans in sight and no indication we would be rescued if an elephant decided to charge. My stomach turned over, and I started to bite my nails. Starting to

THE ELUSIVE BABOON

panic, I reminded Cliff of the elephant stories. He admitted he was not comfortable around so many elephants but assured me we would be fine. Caroline, unaware of the danger, was fascinated and commented on their trunks and tusks and how they flapped their ears, but to me they took on gargantuan proportions, were no longer so fascinating, and seemed much larger than I remembered them in the London Zoo. I'd never seen elephants in the wild before. To be faced with a herd of them was both awe-inspiring and unnerving. With our vehicle as our only defense, I clung to the hope that we could drive faster than an elephant could run, but what if we got wedged in the middle of a bunch of them? Would we be tossed into the air and trampled on like the Germans' mini-bus?

We edged forward. Every so often, one would stop and stare at us through its small deep-set eyes surrounded by wrinkled gray skin, and remembering their unpredictable natures, I went weak at the knees. One moved in our direction, causing fear to course through my veins. I squeaked to Cliff that we needed to gun the Land Rover and get going, but the animal stopped and retreated. Gradually we moved forward, trying not to disturb or provoke them until at last we were through. Suddenly the world returned to normal and the Land Rover took on its usual size. The majestic-looking elephants were wonderful, and I was glad to have seen them but had no wish to repeat the feeling of helplessness and terror in their overwhelming presence.

We also saw herds of buffalo, hartebeest, Uganda kob, reedbucks, water bucks, and graceful gazelles. A profusion of beautiful birds flitted across the sun-drenched landscape like brilliant jewels. We opened the trap-door in the roof of our Land Rover and took turns to look out and see the animals and birds and the surrounding grasslands broken in places by trees. That way, we had great views while still protected by the vehicle. Being an avid bird watcher, Cliff had his binoculars, bird books, and notebooks ready. He made copious notes in tiny script that few apart from him could decipher. No wonder someone had said, "Trying to read your writing literally gives me a pain in the neck!"

After passing through a heavily forested region and brush savannah, we saw another reminder of elephants when we came to an area like a surrealist Dali landscape. Elephants had stripped the bark off hundreds of trees covering several square miles. Subsequent exposure to the sun had bleached the trees completely gray-white. This lifeless, devastated landscape was a ghostly sight. All the trees were gone when we drove through six years later, and I might have believed it was a figment of my imagination except that I had a photograph to prove it.

By mid-day the sun was directly overhead and we were extremely hot. There was no sign of life because everything was resting in the shade. We followed suit, stopped under a tree, and had a picnic of warm corned beef out of a tin, plus bread and fruit. How wonderful our breakfast feast seemed as I ate my rather unappetizing meal and thought, Hey diddly dee, this ain't the life for me. At the same time I knew it was an amazing new adventure. We finished up and got going.

CHAPTER 16

Arua

Huge hippos wallowed in the shallows near the river banks where the road petered out after it sloped down to the Victoria Nile. We pulled up and got out. Coming towards us across the river was a ferry, little more than a flat wooden raft with an outboard motor. Yet this fragile-looking piece of equipment was to take us and the heavy Land Rover to the other side. It drew close, the noise from the engine increased, and suddenly a man in a white shirt and khaki shorts jumped off onto the bank. His colleague threw him a rope, which he hitched to a wooden post to secure the vessel, and then the driver steered the craft towards the landing stage. Close up, the ferry looked even more fragile, but we drove onto it without too much difficulty, and it held.

We exited the vehicle to look out at the scenery and then had much better views of the hippos, whose snouts, huge leathery backs, and small ears stuck up from the mud in which they luxuriated to keep cool. Most of them ignored us, although a few looked back with their piggy eyes as though curious. None came closer, and most looked too comfortable to move. We were the only passengers, and as soon as we were on board, the ferry set off. Small waves formed in our wake as we slowly moved forward with the engine chugging steadily. On the other side the craft docked, was secured, and once it was stable, we climbed into the Land Rover, cranked the engine, and the vehicle struggled slowly up the bank to Paraa, which in Luo means "home of the hippos."

In those days Paraa was a small settlement consisting of a few houses, a small hotel, and a tiny museum. When we visited in 1971, after Idi Amin first came into power but had not yet wreaked destruction in the country, Paraa had become a large tourist center full of Americans. A new hotel with every modern convenience had been built, and a string of Volkswagen buses waited outside to take tourists to see the animals.

This time there was little to see, and we drove through Paraa quickly, but our progress soon became slow and extremely hazardous on a road riddled with potholes. Cliff drove carefully but eventually hit a hole so large that the vehicle stopped dead in its tracks. We both shot up towards the roof, and Caroline was thrown towards the windscreen. Shaken though not badly hurt, we were more concerned about damage to the vehicle, but it started up and we moved forward. For some time we monitored it for signs of problems, but it kept going, and after a while we relaxed and assumed all was well. Nevertheless, the journey continued to be extremely slow and uncomfortable until we saw the Albert Nile ahead of us cutting though the landscape like a glittering golden band in the late afternoon sun. Again the road petered out as we drew up to the banks of the river and crossed by ferry, this time to Pakwach.

In Pakwach, the houses were fascinating and quite different from any we had seen before. Around Kampala they were rectangular and mostly constructed of wattle and daub, though some were of stone or brick. Generally they had simple sloping roofs made of corrugated iron or plain thatch. But at Pakwach the houses were round, with roofs made of dried grasses arranged in layers like a crinoline skirt that came to a point at the top. Though more elab-

THE ELUSIVE BABOON

orate, they reminded me of the Church Missionary Society box in my parents' home, but we didn't linger to look around because we had to reach Arua before dark.

The countryside became drier, and for the first time we saw cotton growing in the fields. Then we headed in a northwesterly direction, began to ascend, and the hazy shapes of the mountains of the Congo appeared on the horizon. We were close. About half an hour before dark, we reached the outskirts of Arua. We'd been traveling on rough roads in intense heat for hours, and by then were hungry and ready to settle in but first had to find the local game warden, Mr. Nkalubo, who was to help us hunt for baboons.

Cliff assured me Arua was a small place, that the police station was probably on the main road, and that someone there would know where to find him. We would soon be settled and could get supper and rest. We found the police station, parked, and went inside. An African in a khaki uniform was leaning back on a chair behind a desk. His half-closed eyes popped open in amazement, and he sat up straight at the unexpected sight of two tall white people accompanied by a small girl with flaxen hair.

As Cliff approached, the man rose to his feet and, through sign language and the name of the game warden, a smile of understanding spread across his face. "Ah! Mr. Nkalubo."

He held up his hand, indicating we should wait, left the room, and a little while later returned with another African, smartly dressed in a khaki uniform, whom he introduced as the district commissioner. Caroline called him the "Deetleman."

Though friendly, the district commissioner was obviously surprised to see us. He spoke good English and wanted to know why we were there before he agreed to take us to the game warden. We thanked the policeman for his help and followed the district commissioner's car for a short distance to the game warden's house, which was surrounded by an extensive area of grass, bushes, and trees. Following his lead, we parked, got out, and trailed after him as he went to the door, knocked, opened it, and entered a shady room with a low ceiling in which lay a bench, a table, a few chairs and some cushions. A houseboy sat on the floor with his back against the wall, reading a book, which he hastily turned over and tucked under a cushion. The game warden was out. Discussion in a language we couldn't understand followed before the district commissioner turned to us and said the houseboy knew where to find the game warden. They would go and get him while we sat and waited. He bowed politely towards Cliff as they left.

I turned to Cliff while Caroline played with her toys. "Did you notice the book the houseboy hid under the cushion was a well-thumbed copy of *Lady Chatterley's Lover?* The English would be a bit advanced, don't you think?"

He laughed. "Yes, I saw. I expect he was picking out the juicy bits. It could be a good way to learn English, I suppose."

"Not sure about that. Magazines seem more appropriate. I wonder how he managed to get his hands on *that*." I sounded like a prim maiden aunt.

"Yes. It's odd to see it here, but I doubt we'll find out where it came from. We can hardly ask."

He was right on both counts. Although initially published in 1929, the unexpurgated version of the book, with its explicit descriptions of sex, had been censored and could not be published openly in the UK until 1960, after the famous Chatterley trial. How had a copy ended up in that remote corner of Uganda?

We continued to wait, getting more hungry and tired. Thirty minutes passed before the houseboy and district commissioner returned accompanied by an Indian of ample proportions in his late thirties. He wore light khaki cotton trousers and a loose-fitting white cotton *kurta* that flowed over his large stomach and ended just below his knees. He was friendly, spoke good English, and said he was Mr. Patel, a friend of the game warden. A minute later, the door opened to admit an African, of slight build and medium height, in khaki shorts and shirt, knee-length socks, and black lace-up boots. He greeted us warmly and said he had been expecting us. He had a pleasant, open face and ready smile. He apologized for his absence when we arrived: He had been called away on business involving poachers. "Welcome to my home. I'm Mr. Nkalubo."

"Uncle Bobo," said Caroline. The man smiled as he looked down at the serious face of the small child who was hugging a teddy bear. To his delight, she smiled back. She'd won a friend. The district commissioner left, waving aside our thanks, and Mr. Nkalubo asked us to sit while he conferred with Mr. Patel in Swahili. Some animated discussion followed before they told us they had decided we should all visit the Indian's house for snacks. I just wanted to eat and go to bed, but Mr. Patel indicated he would be very disappointed if we refused his offer

And so we set off for his house, where his wife and her sister, who were dressed in silk saris, one in bright blue the other in emerald green, flitted around serving tea and round balls of crispy batter with delicious spicy interiors, before disappearing behind a curtain draped across one end of the room. Every so often the small face of a little boy with big brown eyes peeped out from behind the curtain. At nine-thirty, when Caroline was cantankerous and the spicy balls lay heavily on our stomachs, we insisted we must leave and so set off back with our new friend, Mr. Nkalubo.

I was bone tired. But at Mr. Nkalubo's house, the houseboy had prepared dinner in our absence, and we had to sit down to eat yet again. We tried to do the meal of Uganda kob justice, but on top of the spicy balls this was not easy. Mr. Nkalubo told us we could camp in his garden, and we established plans for the following day, pulled the igloo tent from the Land Rover, pumped it up, and finally got to bed around eleven. It wasn't very late, but we had been on the go for about seventeen hours and were exhausted. A day of baboon hunting lay ahead.

CHAPTER 17

A Close Call

I LET OUT A YELL. The houseboy came running from behind the house. In his hand he held a wad of coarse grass that he used to scrub the cooking pots. He stopped and gasped. "It is very bad, madam."

"I know," I wailed. "What are we going to do?"

Cliff and Mr. Nkalubo had headed off early that morning to look for baboons and marshal a team of baboon catchers while I stayed at the camp to tidy up, prepare food, and watch Caroline. I usually kept a strict eye on her, but my attention had wandered briefly, and when I looked up she had disappeared. I dropped everything, ran toward the house, and discovered her covered in soot and carrying a large piece of charcoal that she had taken from the cold ashes at the back, where the houseboy did his cooking.

To my dismay, I found she had also co-opted the services of Mr. Nkalubo's niece, who was staying with him and was about the same age. She too had a piece of charcoal. Together, they were busily drawing all over the porch and the outside of Mr. Nkalubo's house. I stared at the dirty little two-year-olds, their hands and faces streaked in black soot, gazed in horror at the once-clean

THE ELUSIVE BABOON

walls, and shouted at Caroline, "You were *very naughty* to do that!"

She looked back at me with an innocent expression. "Why?"

Sometimes the child exasperated me. Among other things, she had recently filled her father's shoes with water, unraveled a roll of toilet paper all over the bathroom, and emptied a packet of rice in the bed.

"Because you have made a mess on Mr. Nkalubo's nice house and he will be upset and cross."

Her little African companion didn't understand English, but she could tell I was upset. Both children looked subdued and remained quiet as the houseboy and I set to work trying to clean up. He brought more coarse grass and a bucket of water from a rusty tap in the back, and I found a pair of underpants I decided to sacrifice. With the grass and the pants, we managed to rub off most of the mess before we washed the walls thoroughly and took a break. They didn't look too bad, though certainly not as clean as before.

Once the worst was over, the houseboy put down his wad of grasses, looked the two small girls up and down, turned to me with a serious expression on his face, and said, "I think those chaps must wash their hands, madam."

"Those what?" Chaps to me were men.

He pointed at the little girls. "Those two small chaps, madam."

"Ah, yes, of course. The small chaps need a wash."

I hauled the chaps inside and thoroughly scrubbed them while the houseboy went back to his arduous pan scrubbing.

Two things struck me during these exchanges. One was that the houseboy's English was good, and that he certainly understood it well. Maybe *Lady Chatterley's Lover* was not so advanced for him after all. The other was that his expressions had an oddly colonial tone, and reminded me of those in *Up Country Swahili*, the book Cliff had unearthed in the army surplus store in London.

Fortunately, when the men returned later that day, Mr. Nkalubo was not too upset about the house but more concerned about the baboon hunt. They had seen baboons galore but only managed to get one, and this had been at the expense of its life when Mr. Nkalubo shot it. However, they had spoken to a lot of men who enthusiastically agreed to help in a large-scale baboon hunt when we returned to the area. Cliff was disappointed about the baboons but excited because there was a prospect of seeing white rhinos. These massive animals weigh three to eight thousand pounds and are Africa's second largest land mammal. Only the elephant is larger. However, the rhinos had become an endangered species. They had been hunted in colonial times and then poached for their horns, which, when ground to a powder, were thought to have medicinal properties, especially in traditional Asian medicine. As a result, selling rhino horn to that market was highly profitable, and by 1966 the animals' range in Uganda had been reduced to just one small reserve near Arua. Mr. Nkalubo said rangers kept track of their movements to defend them against poachers. If we all wanted to see them, he would make the necessary arrangements. We jumped at the chance.

The next morning we hit the road by five-thirty, heading for the park rangers' camp about twenty miles to the north. On an unfinished section of road, we saw monkeys in the trees at the side and once had to stop because an enormous rhino was strolling down the road ahead of us. By the time we reached the camp, the rangers had left, but a man at the camp gave us directions, and we soon caught up with two rangers on one of the narrow tracks.

When Mr. Nkalubo explained why we were there, the men said they knew where to find the rhinos: They would be glad to take us to them. Climbing into the back of the Land Rover, they directed us over some rough tracks into the bush and told us to park. Everyone got out and, trying to make as little noise as possible, we set off on foot following the rangers into the undergrowth, with one of them carrying Caroline.

Suddenly, the man in front motioned everyone to stop. About twenty yards ahead seven huge rhinos were foraging in a clearing. Their grayish coats were dappled by the light coming through the leaves of the surrounding trees. Small white egrets on their backs were busily picking ticks and mites from

THE ELUSIVE BABOON

their leathery skins.

Two males started to fight, and we watched with baited breath as the huge beasts wrestled with their horns pushing and interlocking as they snorted and shoved at one another. This went on for some time, until one of them backed off and moved away. The other animals took no notice and continued to forage. At one point a large male rhino came over and stared hard in our direction. I was unnerved, but the Africans were not concerned, and Mr. Nkalubo whispered that white rhinos were much less aggressive than black. He said white ones were more likely to run away from us, while black ones were more likely to charge. Thus we held our ground and had a very good view before they moved off and we returned to the Land Rover. One of the Africans was still carrying Caroline, who had fortunately remained silent as we watched, because she had insisted on saying, "I must be quiet," in a loud voice on the track out.

We dropped off the rangers and started back in good spirits towards Arua. It was still early, and our morning had been very productive, with excellent views of the rare white rhinos.

We were discussing our plans for the rest of the day, and had just driven down a steep slope with a drop on either side, when the Land Rover swerved out of control. Cliff managed to bring it to a stop but the steering had broken, probably as a result of having hit the pot hole the first day. It must have cracked and held for a while, but the stress had become too great. Fortunately, we had reached a flat stretch of ground. If it had happened a minute or two earlier, I realized with a shock, we would have plunged off the slope to our deaths.

But we were stuck in the middle of a dry, deserted region. No other vehicles were in sight to help, and Arua, the nearest place with a petrol station or garage, was about twelve miles away. Not anticipating any problems, we had left all our gear at the camp, brought no water, had no means of getting in touch with anyone from the road, and had no air conditioning in the vehicle. It would get very hot before too long, and we had no idea what to do. We didn't speak the language, were not wearing adequate clothing to protect us from the sun, had a small child who would have to be carried if we set out on foot, and for all of us to walk twelve miles in the heat of the day would have been almost impossible. Even if we managed to reach the main road, we had no assurance that any vehicle would pick us up.

But Mr. Nkalubo was still with us. After some discussion, he said he would try to get help while we waited. He knew the area, could speak to the local people, and had contacts back in Arua. With any luck he could make reasonable progress before the sun reached its zenith. There was no alternative, so he climbed down from the vehicle and set off. We watched him trudging along the road, getting smaller and smaller until he disappeared into the distance. All we could do was sit, wait, and hope he could get through.

Hours passed; the sun rose higher and higher until it was beating down relentlessly from a cloudless sky. Behind us the rough, bare track led into the hills from which we had come. Ahead of us, the road stretched like a reddish-brown ribbon as far as the eye could see. Dull browns of flat, parched earth interspersed with occasional small patches of scrub and green vegetation, surrounded us.

No vehicles appeared. The main sound came from the buzzing of insects. We saw only one person, a small, thin woman working in the fields where she was cultivating a small garden by turning the dry soil using a small wooden dowel with a pointed end. She looked quite old, with wrinkled skin that had been dried by exposure to the sun. Her only garment was a small piece of cloth that covered her pubic area. Her uncovered breasts looked shriveled and drooped in small bags. I thought how different she was from my maternal grandmother and her generation. Admittedly the weather was cooler in England and required different clothes, but Grandma always insisted on wearing corsets, vests, and interlock knickers with legs to the knees. For some reason she approved of corsets but not bras. Maybe this was because bras, or "brassieres" as she always referred to them, had not been around when she was a young woman. The first modern brassiere to receive a patent was the one

THE ELUSIVE BABOON

invented in 1913 by a New York socialite, Mary Phelps Jacob, when my grandma was in her twenties. Why exactly she resisted them, I wasn't sure. What I *did* know was that the thought of working outside almost naked would have shocked her sensibilities to the core.

With these thoughts drifting through my mind, the sun continued to beat down, the temperature rose to over ninety degrees, and we were getting very thirsty. Occasionally one of us dozed while Caroline slept. The roof of the Land Rover was our sole protection, and although it kept off the sun's direct rays, the metal absorbed the heat. The windows were open, but there was hardly any breeze. It was like being in an oven. Our clothes and hair clung to our sweaty bodies. Time passed slowly, and we began to wonder whether we had been abandoned but had no idea what to do and whether we would survive the awful situation intact. If no one came, we would have to set off on foot, carrying Caroline much of the way, but not until the heat abated, and then who knew what danger we would encounter. I was three months pregnant, and the baby seemed secure, but would this put too much strain on my body? I hardly dare think about it.

Six exhausting hours passed with no food or water, and then the welcome shape of an oncoming vehicle appeared in the distance. Gradually it grew bigger and, as it came into full view, we saw Mr. Nkalubo sitting in front of a tow truck with a driver. He hadn't let us down, and I was so pleased and relieved to see him that I didn't know whether to laugh or cry. I took a deep breath and tried to contain my emotions as I had learned to do as a child when I dug my nails into my palms and bit my lip. The vehicle drew to a halt; Mr. Nkalubo jumped down, walked over to the Land Rover, and told us his story.

After he left us, he'd had to walk several miles to the main road. This had taken some time. Once there, he managed to catch a bus that had taken him at a slow pace into Arua, where he had to find help and persuade someone to come all the way out to get us. It had not been easy. He had saved us, and we couldn't find enough words to tell him how grateful we were.

The truck driver was a cheerful soul who hitched us to his vehicle and set off at a slow, steady pace. My spirits began to lift, and the terror of the last few hours started to subside. When we eventually joined the main road, we saw people staggering along with heavy loads on their heads such as cassava bundled like firewood. Women carried small and large water containers, and even small children struggled along with bunches of bananas. In addition to loads on their heads, some women had babies strapped to their backs, some in cloth slings similar to ones we had seen in Kampala, but more often strapped to the mothers' backs and carried in woven containers like envelopes that had been sealed on a long and short side, leaving the other two sides open. Some envelopes were in a plain cream weave, but others had a brown zigzag pattern woven into them and looked like small works of art. The short, sealed side of the envelope fitted over the baby's head like a hood; the long side hung down to cover its back and protect it from the sun, while the open sides were next to the mother's back. By carrying her baby in this way, the mother's arms were free to work while her baby was rocked by her movements and comforted by the closeness of her body. Small faces with large brown eyes peeped out from the sides of the hoods.

Bicycles transported heavy goods and often were loaded with tables, chairs, sacks, or other people hanging on precariously behind. Occasionally a bus went by filled to overflowing with people hanging on at the door and sides. Men clung to the backs of lorries. Cars were rare. Those we saw were generally Peugeots filled to capacity with at least six to eight people. Drivers sped recklessly along the bumpy *murrum* surface, kicking up dust that covered the passers-by. All of this drove home the difficulties we would have faced had we abandoned the Land Rover and tried to walk or hitch a lift. I couldn't begin to imagine what might have happened if we had not had the great good fortune to have Mr. Nkalubo with us.

Compared with the Baganda around Kampala, the people of the West Nile had darker skins, less broad features, and were much thinner. Some older women followed tradition and didn't cover their breasts. Sometimes they wore little more than a pelvic cloth, like the woman we had seen in the field; some wore only a small bunch of leaves to cover their pelvic area. However, many younger women had adopted a more Western style of dress, and many older women wore the voluminous national costume like that of Buganda. You could see the area was changing.

Finally, we reached Arua and took the car to be repaired. There was no repair shop or garage in any traditional sense. This one consisted of a small shed and an open stretch of ground where a varied collection of used spare parts from vehicles beyond repair were displayed. The African mechanics were inventive and resourceful, and used any salvageable parts to make other vehicles roadworthy. They would have to salvage pieces from other vehicles to repair our Land Rover. The Rover posed a difficult problem, but the mechanics didn't complain. They were more inclined to view the job of fixing it as a challenge and a welcome excuse for people to get together, chat, discuss the problem, help out, and generally socialize. No one said, "Sorry, we can't deal with that," or "You have to have a special part, and only a new one will do." Even when faced with a time-consuming job as ours turned out to be, they set to work creatively to repair it.

The breakdown cost us time, but this was a minor drawback compared to what could have happened. Sadly, the trip was not successful in terms of catching baboons, but we had seen the rare White Rhinos, and made a new, valued, and kind friend in Mr. Nkalubo. He assured us he would arrange for a team of men to help catch baboons when we returned. Meanwhile, we had to wait for the vehicle to be repaired before we faced the long journey back to Kampala.

CHAPTER 18

Buffaloes at Night

As soon as the vehicle was ready, we set off in the early afternoon of December 22 knowing we would have to camp overnight if we were to reach Kampala on the 23rd as planned. The breakdown had unnerved me more than I cared to admit. We were taking the same route back, and I was beset with fears. What if it broke down again on the rough roads? What if it broke down in the middle of Murchison Park, near the gigantic elephants? What would we do with no help nearby? But the Land Rover seemed to be taking the roads in its stride. By early evening we reached the north gates of Murchison Park, where we found a cluster of round metal huts occupied by park rangers and their families. I picked up a brochure we had brought and turned to the information section.

"According to this, we're at the camping ground we talked about before setting off. It hardly looks like a mecca for tourists. In fact, I doubt anyone has ever pitched a tent here."

"I know. It *does* look a bit primitive, but we'll have to stay, because it won't be long before dark and the park closes. We couldn't camp in there anyway.

THE ELUSIVE BABOON

This is the place we agreed on before we set off, and there's nowhere else. It shouldn't be too bad. At least there are people around."

There were no camping facilities, but it was our only option. We pulled up on a rough patch of ground before setting up camp on some grass at a little distance from the Land Rover, where we were within easy view of the ranger station. Men, women, and children came out of the huts and stared as though surprised to see us, but they appeared friendly because some waved.

We rigged up the tent and sat outside to watch chickens scratching around in the dust and women cooking with charcoal in front of the metal homes. When the sun started to go down, we cooked our own evening meal while Caroline ventured off to make friends with some children living in the huts. She found a small boy, and the two of them ran around squealing happily. On reflection, it had been a successful day. The Land Rover had been fixed at a reasonable price; we had come to the park in the early evening, at an ideal time to see the animals, and had some wonderful views of Uganda kob bounding through the grass in the golden light of the setting sun. Everything appeared peaceful and pleasant. As darkness descended, we lit the paraffin lamp, which gave enough light to read by before settling down for the night. We looked forward to being back in our own comfortable beds the next day.

At around one or two in the morning, Caroline started to cry for no apparent reason. Maybe she had had a dream, for she quieted down when I moved close and held her. After a while I needed to relieve my full bladder and grabbed a torch, stepped outside, and wandered off to find an appropriate tree or bush, for we had no other facilities. I was squatting out there when I heard the sound of something moving in the distance, but it was very dark and I could see nothing. Nevertheless, the noise was disconcerting. It was getting louder and appeared to be advancing rapidly in my direction. I had no idea what it was but didn't wait to find out and scampered back to the tent and the protection of the sleeping bag, where I tried to settle down. But the noise increased and sounded like something very heavy was tramping over the ground.

It was coming in our direction and, because I had no intention of experiencing this particular thrill on my own, I nervously poked Cliff, who grumbled but woke up. "What's the matter?"

"Listen," I whispered, "there's something out there. What's that noise?"

It was getting louder and louder. Grass was being ripped up, and it sounded as if hundreds of people were tearing strips of cloth.

"Buffalo, I expect," he whispered and sat up.

"Buffalo! Oh, my God—what are we going to do?"

"Nothing. Keep quiet. If buffalo are anything like cows, they'll come to investigate any noise. They may investigate the tent anyway."

This did nothing to reassure me. Had we been surrounded by cows, I would have been frightened. I had never been the one to lead the charge through fields of cows on geography trips when younger. But buffalo struck me as ten times worse. They are not animals you trifle with. With thick legs and stocky bodies covered in black and dark brown coats, adult males can weigh well over a thousand pounds. They have large horns that curve out and up on either side of their heads. Like elephants, they have unpredictable natures, and when roused buffalos may attack and gore people to death. We were in the path of goodness knew how many, and they were about to descend on us. I felt trapped and helpless. "We can't just *lie* here. They may trample on us. We could get killed."

"Well, you come up with a better solution."

The only thing I could think of was to shelter in the Land Rover, but it was parked some distance from the tent, and by then the animals were too close for us to reach it without encountering them. We were stuck. Our canvas tent was no match for the bulk and horns of even one buffalo if it chose to charge. As they grazed and came closer, the night air filled with the noise of their heavy breathing, and every so often one would break wind, which sounded like thunder overhead. Once we heard the sound of an animal pawing the ground, and I wondered if this meant it would charge and we should be

THE ELUSIVE BABOON

saying our last prayers. I was terrified that Caroline would wake again and start to cry, so I lay close, held her to me, and almost smothered the poor little thing in my determination to let no sound escape. We stayed in petrified silence as the cacophony of noises continued.

For an hour we listened, hardly daring to breath. It seemed like an eternity. And then the noises began to subside, becoming lower and lower, until we could hear them no more. The buffalo had left almost as quickly as they came. Even then we daren't look out for fear a stray animal had stayed behind.

Eventually I took a deep breath. I was still clinging to Caroline and, when I began to relax my grip, discovered I was shaking all over. Cliff settled down, but I didn't sleep any more that night, and dawn was a welcome sight.

In the morning I was first up and peered through the tent flap. A smell of cut grass and smoke from the settlement filled the air. Enormous animal droppings covered the ground only about two feet away and certainly too close for comfort. Stiff from being so tense during the night, I emerged with some difficulty, walked a little, and then started to stretch. As I extended my arms to the heavens, there was a terrible clanging as though someone were forcibly striking a metal gong with a large hammer. Men began to appear from all directions, carrying hoes and what appeared to be weapons. My nerves were still upset, I was feeling disoriented, and I didn't know what was happening but, after the buffalo, was prepared to think the worst. Maybe this was a lynch mob.

Like the man in the circus who is shot from the barrel of a cannon, I dove between the flaps of the tent and landed on top of Cliff, who by then was awake but still in his sleeping bag. He uttered a startled gasp. "What the hell do you think you're *doing*, Jen? You almost winded me!" He struggled to get his breath back.

"Sorry, but did you hear that noise? They're all out there, ready to attack."

"What on earth are you talking about?"

"I just went out for a stretch, and suddenly there was this terrible clang. And they started to appear with hoes and things, and then they lined up."

"Who started to appear?"

"These *men*. They look fierce to me. I'm scared."

"I suppose you didn't scare me by landing on top of me like that out of nowhere?"

His sarcasm was lost on me. "Oh, nothing scares you!" I grumbled. "I'm telling you there are men out there all lined up with weapons, ready to attack. No wonder they weren't very friendly last night."

His eyes were wide with amazement and disbelief. "They were *not* unfriendly, you said so yourself at the time, and they are *not* carrying weapons. Those men are getting ready to head off for the day to guard the park. Probably they have a roll call first, and that's what it's all about."

I was defensive. "I suppose you think I invented the buffaloes, too."

"Of course not. That was totally different. I can understand you being upset by the buffaloes. They were real, but gangs of attacking men are purely a figment of your vivid imagination."

"But that's the point. No one knows where danger lurks."

He threw up his hands in exasperation at this illogical statement: "Of course they don't, but most people don't go looking for danger and making things up."

He looked at my serious face and began to chuckle: "Whatever will you think of next?"

"Silly Mummy," said Caroline who had been listening to the exchange and had to stick her little oar in.

"I'm glad you're both so amused. Anyway, why are you encouraging her?"

His jaw dropped open at the effrontery of this accusation, but before he had time to sputter a response, I turned toward the front flap of the tent and peered out. Sure enough, the men had assembled in front of the main hut from which the clanging came. As I watched, a roll call was taken, and they were dispatched to work. I began to feel very foolish. "It looks as if you were right," I admitted grudgingly.

He grinned. "Yes, I usually am." I batted him, and he continued, "Still, you're always good for some entertainment. Never a dull moment," he reflected as he got up and started to dress. "Poor old Jen. I know you've not been feeling well, but you should try not to let things get to you so much."

I had nothing more to say on the matter. My unbalanced hormones, lack of sleep, overactive imagination, and the buffaloes had played tricks on my nerves. What seemed like a light and benevolent environment the previous day, now seemed dark and menacing. I wanted to get back.

Fortunately, the rest of the journey took place without incident. We reached Kampala on the 23rd in time to do some Christmas shopping the next day.

That year, for the first time, we were going to experience Christmas in the heat. And then Cliff was taking a short break from baboon hunting to look for bush babies.

CHAPTER 19

Bush Babies

ALAN WAS SETTING UP A NOCTARIUM at Makerere so he could manipulate the light to reverse night and day. He planned to put bush-babies in it. The idea was that people could observe their behavior more easily, because these small nocturnal creatures were almost impossible to watch in the wild. However, he needed animals to stock the noctarium. Cliff said he would help catch some.

Bush-babies are delightful-looking little creatures with large, round, limpid, sherry-colored eyes that almost fill their tiny faces. Five to eight inches in length, and about six to eight ounces in weight, they have an appealing way of holding food in their small paws while nibbling it in a dainty fashion. With large batlike ears sticking up from the tops of their heads, small furry gray bodies, and long furry tails, they look like tiny cuddly toys. It's easy to get sucked in by their looks, but they are *not* ideal pets. They sleep during the day and kick up a rumpus at night when they race around and let out cries like a baby in distress.

The first attempt to catch some took place soon after we arrived when,

THE ELUSIVE BABOON

one evening, Cliff and Alan drove about twenty miles to Entebbe to pick up eight muscular young African men identified by a contact at the Virus Research Institute as "catchers of bush-babies." The catchers piled into the Land Rover, and the one who spoke English said they had to drive further. After an unexpected twenty miles, they reached the forest and were instructed to stop. The catchers jumped out, headed to a spot where they cleared a patch of ground, and the spokesman announced they had to shake the trees so the bush-babies would come out of their nests, fall to the ground, and be grabbed. Cliff and Alan told them to go ahead.

The leader issued a command, the men leapt forward, vigorously shook the trees, and a series of plops followed when some animals fell to the ground. But the men were too slow. Before they could pounce, the animals beat a hasty retreat. Ultimately the total yield from eight strong men was two animals about twice the size of a house mouse. The whole thing was a farce, with the catchers jumping, pouncing, and leaping to no avail while enjoying themselves enormously.

Cliff and Alan were not amused. Catching bush-babies obviously required another strategy, but the ones we personally acquired came purely by chance.

After December's unsuccessful baboon hunt in the West Nile, Cliff decided to take a break. He had heard some geologists at Makerere planned to visit a fossil site recently discovered at Bukwa, located on the volcanic cone of Mount Elgon on the Kenya–Uganda border. He was interested in fossils. I wanted to see a different part of Uganda while I was still mobile. We decided to join them.

The air was still and the sky a clear, vivid blue when we set off early one morning during the third week in January, joined the rest of the party in Kampala, and headed east in convoy with Alan in the lead. Two passengers sat on our back seat while Caroline sat between us in front. There was little else on the road, and we made good progress. After about forty-five miles we paused near Jinja. Here for the first time I saw tall yellow-green stalks of sugar cane with grass-like leaves sprouting from the tops, and the thick, densely packed, light-olive-green leaves of tea growing in fields that stretched like an enormous lawn into the distance. We also saw the impressive concrete dam at Owen Falls, built for hydroelectric power and spanning the White Nile not far from Lake Victoria. Water forced its way through several sluice gates and shot white foam several feet into the air in front of us.

We drove on until the large volcanic plug marking Tororo appeared. Thirty more miles brought us to Mbale, at the foot of Mount Elgon. We didn't stop but drove through town, left the main road, and began our ascent along a narrow, winding dirt track cut into the side of the mountain. This rapidly deteriorated into ruts and rocks, causing the Land Rover to rattle and creak ominously. Our equipment was flung around in the back, I hung onto Caroline, and the Land Rover was forced to a crawl. But we saw breathtaking views. At times we looked down on seemingly endless plains onto which puffy white clouds that dotted the sky cast dark liquid shadows like vast lakes. Small volcanic cones, looking like golden-brown mounds in the sun, covered other areas. Bordering all of this, mountains formed inky-blue silhouettes against the blue-grey horizon.

About forty miles from Mbale, we stopped for a late lunch in Kapchorwa, the main administrative center in the Kapchorwa district, then continued at our snail's pace along the rough, winding mountain road, continuing to be jostled around, until we finally reached our destination just before sunset. There we joined the rest of the party, put up the tent, had a quick meal and, after a very long day, retired for the night.

Early the next morning, we drove about two miles along a narrow rough track to the geological site where prickly bushes and plants with small, thick, shiny green leaves covered the dry ground in a rather inhospitable landscape. We erected a canvas shade near the Land Rover, and the men started to search for fossils. Excitement broke out when they found fossilized leaves and shells, while Cliff caused a stir by unearthing a tooth from a Dinotherium, an extinct animal distantly related to elephants. I felt like a celebrity, because the Africans who lived nearby had had little contact with Europeans apart from those in the missionary schools. Hordes of people followed us, and each day an audience settled down to watch from the boundaries of the site.

Caroline immediately started to make friends with the local children. Her white skin, straight blond hair, and pink-and-blue-striped cotton dress, provided a stark contrast to the African children with their close-cropped curly black hair and faded, well-worn clothing. Some wore brightly colored beads around their navels, ankles, and wrists, and these looked well against their dark skins. But I was disturbed to see that many children had runny noses and a yellow discharge oozing from their eyes which, together with terrible sores on their arms, attracted flies in thick, black, crusty clusters. I shrank at the thought of touching them, but the children didn't seem much bothered by the flies, as though resigned to their presence.

Some children had protruding bellies as well, and I dreaded to think what parasites might lurk in there, for they were skinny otherwise. One child limped on crutches. All looked as though they desperately needed medical help, but all I could offer was some antiseptic cream, which the mothers gladly accepted. They begged for more, but I had no more to give. In any case, this was merely a Band-Aid solution to their dreadful problems. Before we set off, it had never occurred to me that Caroline might be exposed to children with infectious diseases. Now I began to question the wisdom in coming and monitored her carefully.

The children were fascinated by our daughter's toys, especially a big white doll with long, curly blond hair. "This is Gynneth. Say hello," she told them.

Having no idea what she was saying, they laughed nervously. Apparently they had never seen a doll because they kept pointing to it and whispering to one another with frowns on their faces. Finally they seemed to realize what it was, went into fits of giggles, and were thrilled when Caroline let them hold it. A teddy bear, toy cat, and dog also intrigued them, because they had never before seen soft toys. Small children generally spent time playing with one another at made-up games. Older children looked after the younger ones, helped around the house, learned the roles their mothers and fathers played in society, and prepared to do the same.

The local chief soon appeared. He carried a gnarled stick. In contrast to the other men, who wore shorts, shirts, and no hats, the chief wore an old fedora like Bogart's in *Casablanca*, and a black cotton raincoat that extended to his calves to cover his thin frame. His weather-beaten face looked like old leather, and from the end of his chin a pure-white trapezoid-shaped beard, about three inches long with the smaller side at the bottom, stuck out like the bristles of a paint brush. It was so precisely trimmed that it looked artificial and glued on. His earlobes were extraordinary and extended for about two inches due to huge pierced holes that were weighed down and further elongated by heavy earrings. Each earring was composed of a thick band of copper, about quarter of an inch wide, that threaded through the hole in the lobe and hung down to form a loop at the bottom. From this dangled two copper bells each about half an inch long.

On the surface the chief was a pleasant and accommodating old man, but I observed him carefully and noticed that his glinting dark-brown eyes noticed everything that was going on from under his furrowed brow. I concluded his appearance cloaked a calculating mind. This was confirmed when he walked purposefully towards us on the second day, pulled some rough amethyst crystals from his pocket, and told us we could have them for a thousand shillings. This was a ridiculous sum. We refused to spend more than fifteen.

He frowned in annoyance but was not put off and immediately claimed that, because the land on which we were camping was his, we must pay him to stay there. Fearing there might be trouble, I watched apprehensively while he and a member of our team negotiated until they struck a curious bargain. It turned out the chief liked money, tobacco, and beer. He agreed that we would buy three guinea fowl for five shillings each, provide him with cigarettes and beer, and give his son a lift into Kampala to catch the bus to Bombo, where he was going to school. These promises seemed to satisfy him, but I was surprised he would trust his son, who looked to be about twelve, with a bunch of strangers on the long journey back.

Meanwhile, a crowd of men had gathered around Alan. He told them he wanted to collect some live bush-babies and would pay for them. We were about to depart when two bush-babies were brought to him in bags. Once they knew money was in the offing, the men wanted to know when we would be back to collect more. We promised that, one day, someone would return. (The excavation itself was very successful. In 1969, Alan published the results in the journal *Nature*.)

Several months passed, and I had forgotten about the bush-babies until, one evening, Cliff's disheveled figure came through the back door, struggling with a muddy sack. He gently lowered it to the floor. His khaki shorts and shirt were filthy, his boots were covered in mud, but he had a big smile on his face.

"Whatever's that?" I gasped as the sack started to move.

"Bush-babies."

"What?"

"Bush-babies, and we're going to keep them. We've been trying to get hold of some for ages. These were a real stroke of luck!"

"Well, yes, but how many have you got?"

"Only eight. They can go in the spare bedroom. I have some cages there."

"Spare bedroom? Only eight? That's one heck of lot. We can't possibly . . . ."

"We've no alternative for tonight. Some can go into Makerere tomorrow, but I want to keep a few. Now, they need to rest, and so do I. We've had a terribly long journey."

He was right, we *had* been looking for bush babies for a long time, but I'd never planned for eight to take up lodging in the spare bedroom. I was too tired to argue, so they stayed for the night.

The next day I found out what had happened.

Cliff had been on a four-day trip to Elgon with Don Smith, a Canadian bat specialist who'd come to Uganda because of its great variety of bats. Cliff wanted to go back to the fossil site we had visited with the geologists in January, and Don hoped to catch some bats on the trip. When the men reached the mountain track to Bukwa, the rains had been heavy, causing the Land Rover to skid into a ditch several times on the rough surface. But locals appeared as if from nowhere to help push and pull the vehicle, laughing good-naturedly as they got the four wheels back on track.

Don used mist nets to capture some bats before the men visited the geological site. As they were leaving some young men, carrying baskets and hampers, appeared saying they had bush-babies for the *bwanas*. Cliff had forgotten the promise he'd made on the previous visit, but the Africans had not forgotten him. Now they demanded payment.

However, Cliff and Don had gone to see the fossil site and collect bats. They had no containers for bush-babies. They also had a limited amount of cash and had to figure out how to pay for their unexpected windfall. The small crowd started to agitate for money and surrounded the two white men, who towered over them. Good working relations with the locals were important. Something had to be done.

Much haggling followed until a price was agreed on, but then the men had to get the bush-babies back to Kampala without appropriate containers. They decided to give them plenty of bananas and put them in a sack used to provide traction for the tires when the Land Rover skidded in the mud. Holes in the sacking would provide air. And so the animals were carefully transferred from the baskets and hampers before the sack of wriggling bodies was loaded into the back of the vehicle.

But more complications followed when the Land Rover had to be driven slowly and with great care so as not to jolt the bush-babies too much going

THE ELUSIVE BABOON

down the extremely rough mountain road.

They camped overnight and, on the day they were due home, stopped for a late breakfast at a hotel in Tororo. Their journey had been slow and arduous, and they decided to take the bush-babies out of the vehicle to give them a rest and some fresher air through the holes in the sacking. They carried the sack into the hotel and set it down at the side of their table. During their first course, it lay there quietly.

But the bush-babies started to come to life, and the sack began to move.

A waiter was standing against a wall, scanning the dining room. Cliff looked up, saw the man's eyes grow wide with curiosity when they reached the heaving sack, and decided to explain. He called him over. With his eyes riveted on the sack the waiter cautiously stepped forward.

Cliff, ever logical, said, "Don't worry. We've been to Mount Elgon. We were given some bush-babies, and they are in the sack."

"Yes, sir," stammered the waiter.

"We're taking them back to Kampala."

"Yes, sir!"

The sack moved again, and the waiter thought a moment, then said cheerfully as he cleared the dishes, "Good idea, sir. Give them a proper education!"

The men beat a hasty retreat before the local police arrived to arrest the "baby snatchers"!

In the end, Alan took five bush-babies, and we kept three. They were pretty creatures, but eventually their nightly noise was too much, and we gave them to the noctarium. Meanwhile, baboon trapping continued to be a challenge.

CHAPTER 20

Camp Budongo

BABOONS LIVED IN ABUNDANCE WITHIN, and on the outskirts of, the Budongo Forest. Cliff often went there alone, paying his first visit at the beginning of February. On that occasion, he returned with no baboon blood but came into the house, fished something wrapped in a blood-soaked tea-towel from his knap-sack, and handed it to me, saying, "Take this. It's a hunk of buffalo meat from an animal John Bindernagel shot."

I hadn't met John but knew he was a Canadian wild life biologist who studied buffaloes. I had never eaten buffalo but took the dripping package, discarded the wrapping, and put the meat in the refrigerator to make into a stew.

Distracted by the slab of buffalo meat, I hadn't noticed Cliff was sick. But when he leaned against the fridge, I saw beads of sweat forming on his forehead and trickling down his cheeks. His skin burned to the touch, but he said he was freezing. Then he staggered to the bedroom and collapsed on the bed. He said he had willed himself to keep going on the way back and, before setting off, had spent hours lying in the sun, trying to get warm and muster up the energy to drive. It sounded like sheer hell. Good luck, and his dogged persistence in the face of adversity, had seen him through.

I was appalled when I discovered he had eaten some corned beef given to him after it had been left outside in an open tin can in an area infested with disease-carrying flies. He was usually far more careful, but I suppose he was foolishly trying to be polite.

For days he was laid low with severe intestinal problems, and the effects lasted much longer. As usual, he stubbornly refused to see a doctor, and I began to worry about what would happen to me, Caroline, and the unborn child if he got really ill and didn't recover.

But Cliff was intent on his work. He felt time was passing, and he still needed to capture baboons. As soon as he began to improve, he agitated to go back to Budongo, where he had left a baboon trap. If all went according to plan, the baboons would get used to it, go inside for bait, and could then be trapped, sedated, and bled. Nothing could dissuade him, so I decided we should all go. I refused to stay home and worry about him dropping dead in camp. The fear of being a widow with two young children, one of them not yet born, outweighed any fears about what might happen to me in my physically vulnerable state. I argued that, if Caroline and I went, I could make sure the food was properly stored and cooked to avoid further problems.

I had also never been to Budongo and looked forward to a new experience.

On February 21, I was buying food in the Kampala open-air market in preparation for the trip when a terrific hullabaloo broke out. Men rushed from all directions, hollering, whooping, and flapping their hands up and down on their mouths and making sounds like *woo-woo-woo*. The mob headed off down the road, and I asked the young man carrying my basket what had happened. He said a man had stolen from someone in the market. The crowd was out to get the thief and would beat and possibly kill him. It all happened so quickly that the incident didn't sink in until the fleeing bodies disappeared into the distance. This was the first time I had seen first-hand how extreme violence could erupt rapidly and realized you could easily get caught unwittingly in the middle. That day I was glad to get back in the car and return home.

Almost four months had passed since we arrived in Uganda; we kept hearing rumors of political trouble, but nothing serious had occurred. However, on February 22, the political problems that had been simmering began to bubble to the surface.

We were eating dinner when Rikki rushed in. Her hair was askew, and she seemed out of breath. I put down my knife and fork. Cliff stopped eating. "Sit down. What's the matter?"

The words tumbled out of her mouth. "Can't stop, but I know you're planning a trip to Budongo. I heard five government ministers were arrested after they accused Obote and several other people, among them a Colonel Idi Amin, of selling gold for their own personal gain. The gold came from the Congo [now the Democratic Republic of the Congo] and was to be *stored* in Uganda, not stolen."

". . .How serious is it?"

"No one seems to know. But people think fighting could break out, and it might be dangerous to travel."

A wave of fear spread through me at this unexpected and unwelcome news.

"I have to go get dinner," she said, "but I thought I should tell you."

"Yes, of course. Thanks."

She departed, and I turned to Cliff: "Now what?" I wondered if we would have to cancel our trip. Did this mean people would become violent and start to kill, as I had seen in the market? Could we be caught in the middle of disturbances?

We planned to leave the following day. Our gear was ready and our provisions of rice, spaghetti, tea, bananas, and cans of meat were assembled, together with vegetables and bananas, but after this alarming news, we daren't set off.

We needed a reliable source of information, not rumors, to make a decision. Cliff thought the British High Commission would be our best bet, so we drove to Kampala the following morning, parked outside the Commission, entered its cool interior, and headed to the information room. A balding man of about forty, with a paunch, was sitting there in his shirt sleeves, behind a mahogany desk. He looked up from the newspaper in which he was engrossed, hailed us, indicated we should sit down, put down his paper, and listened to our concerns.

"What would you advise us to do?"

He leaned back in his chair, stretched his arms in front of him, and drew a deep breath. Then he put his hands behind his head, exhaled through his mouth while puffing out his lips, and quite nonchalantly said to Cliff, "Not to worry, old chap. If the buggers do anything, you'll probably be better off in Budongo than here."

His drawl sounded very British colonial. He didn't seem in the least concerned, and we began to feel rather silly for having asked the question.

"It looks as if we can go then?"

"Don't see why not." He went back to his paper.

Once out of earshot, we discussed what he said. "It sounded as though something *could* happen, didn't it, Cliff?"

"I suppose so, but it seems there's no point in waiting around. He's probably right. If there's trouble, it's more likely to be here than in the forest."

"Yes, but what happens if we get stuck there?"

"Probably won't."

"I hope you're right. We can't stay in the forest indefinitely. We haven't enough food."

"Don't look for trouble, Jen. I think we'll be fine. We can't sit around guessing."

The next day, we packed the Land Rover and headed north. For much of the journey the rain came down in torrents, turning the roads into mud but cooling things down to provide relief from the heat. A damp, earthy smell rose into the air. On approaching Masindi, the vegetation thinned out and we came into an area where baobab trees grew at intervals over the landscape. They were very tall, with a bulge at the top from which sprouted a few small branches with leaves, rather like a long neck morphing into a square bald head with long, thin hairs attached.

After Masindi, we headed west towards Budongo until we turned off into the forest along a track that led to the sawmills. We stopped at the office to see the manager, Mr. Knight, to catch up on the latest news, but he was not there and we continued. Tire tracks from other vehicles marked the way in the muddy undergrowth. Overhead, the dense canopy of the semi-deciduous forest, together with magnificent old mahoganies that soared skywards, sheltered us from the sun.

After a mile or two of slow driving, the vegetation began to thin out, and we reached the far edge of the forest. An open area of scrub lay ahead. Small trees grew out of tall grasses, and the land was covered in conical, earthy-red ant hills about four to six feet tall. Two female baboons, one with a baby, were sitting on one of them. We stopped briefly to look; they stared back dispassionately, without moving, before we drove on along a track skirting the forest, crossed a small stream, turned left, and came upon a clearing set on a knoll where the grass was fairly short and sparse. We had arrived.

The Game Department had set up a small permanent camp for hunters employed to shoot elephants that ventured beyond a certain area, because the large elephant population caused extensive damage to the trees. The rangers had just come back from a day's work. We introduced ourselves and found they were a friendly bunch whom we were glad to have close by in case of danger. We parked the Land Rover, inflated our igloo tent, and located a flat space for a wood fire.

Our meals were eaten under a shade made of an orange canvas sheet attached by cord to the trees and to posts we drove into the ground. We gathered sticks and logs from the surrounding area to make a fire and over this cooked food in small pans from our camping set.

One day the hunters gave us a hunk of buffalo, which I cooked as a stew in the embers of the fire. Another day, they gave us wart-hog, which was like pork but tough. The metal cabin trunk in which we carried gear became our table, our seats stones or round logs. Caroline acquired her own special seat consisting of an elephant's jaw that Cliff discovered in his wanderings. She fitted right in, and we transported the jaw back to Katalemwa.

Camping was difficult. There was nowhere to wash, and toilet facilities were extremely primitive, consisting of a hole in the ground. At noon the heat was overwhelming, for the sun beat down and the shade provided little relief. I was nearly six months pregnant, had put on too much weight, and was feeling very tired. My body trundled along behind my protruding stomach; my legs felt like lead weights. Walking in the heat was like dragging heavy objects through a furnace.

I knew it would be my last camping trip to that particular spot, and I wasn't sorry. Everywhere was fly-infested where raw meat was disposed of by the hunters, and I was put off by the vultures that perched in some scrubby trees at the edge of the camp. Tatty-looking brown-black feathers covered their large bodies. With scrawny necks, bald heads, and beady eyes, I found them unattractive birds. Clearly, they had their place in Nature but there was something sinister about the way in which they waited to pounce on the waste and innards of dead creatures that few creatures but hyenas wanted. Sometimes a rather ill-tempered marabou stork accompanied them in their scavenging. This enormous bird, at least three feet tall, had a huge wing span of seven to nine feet. Black feathers on its back looked like an undertaker's coat as it strutted around on its long, skinny legs, scavenging among the waste. Like the vultures, its head was bare so it could bury itself in rotting carcasses without the bits of dead meat catching in its feathers.

At night we heard hyenas rooting through the garbage and the unnerving noise of them cracking bones in their ferociously strong jaws. One night I woke to hear lions roaring close by and remembered the man-eaters of Tsavo. When I thought of people torn apart by huge teeth and large claws I lay still, trying to quell my terror. I would have hated to see a lion shot but was reassured to know that men with guns were close by to protect us in an emergency.

One night a severe thunderstorm raged as though an angry god was showing his great power and the ultimate weakness of man in dealing with his wrath. Lightning lit up the inside of the tent, intense claps of thunder shook it, and water came in at the seams when the torrential rain poured down and the wind almost lifted it off the floor. The noise was tremendous, especially when the storm was directly overhead. I held Caroline tightly while Cliff assured me all would be well because he had experienced these storms before. But I felt completely powerless, knowing we could do nothing to stop the onslaught. Through all the battering, the valiant tent held its ground, and gradually the storm moved on, leaving us with a strange sense of quiet broken by the rumbling and echoing of thunder from the distant hills for a long time afterwards.

The narrow, shallow Waisoke River ran out of the forest close by and became our only source of water when our fresh supply ran out. It looked so murky that Cliff said when he was a boy, he changed the water in his tadpole tank when the water was that color. Desperate for a drink, we boiled it thoroughly to kill any bacteria and parasites, added tea and powdered milk, drank it, and came to no harm. But I became aware that fresh water was an extremely precious commodity and, with all the disease carrying flies around, knew it was critical to keep things covered, clean, and thoroughly cooked. I also realized that, even if I were not pregnant, I lacked Cliff's fortitude and ability to deal calmly with the kind of adversity we faced in surroundings like this.

Each day, the hunters went out with their guns and returned to the camp in the evening. Then they lit a fire on which they cooked their meal, and they sat around the fire, chatting and strumming on musical instruments, after eating. At the end of a hot, wearing day the sound was pleasant and relaxing. Caroline played with her toys and went over to the hunters at the campsite, unconcerned that she did not speak their language. She watched them prepare food and hang meat to dry on tree branches, and she sat quite comfortably with them around the fire. Maybe she understood more than we realized, for like many children of that age she was picking up languages quickly and constantly acquired new words. The hunters in turn seemed fascinated by a small white girl with flaxen hair so unlike their own black curly locks. They watched, pointed, and commented to one another as she wandered around, dressed in a pink-and-white checked dress and dragging a teddy bear or trailing her favorite Richard Scarry book. She was a remarkably self-sufficient two-year-old who didn't ask many questions about the hunters but seemed to gather all the information she wanted for herself and accepted them as part of the camp.

By evening, the temperatures had dropped, and we could appreciate more readily the surrounding countryside. At that time of day we relaxed on the top of the knoll, watched the hunters, and enjoyed the panoramic view across the different greens of the forest ranging from a pale grass color to dark olive. Beyond this, the browns and dusty yellows of the open plains were dotted with small, spiny bushes. The vast skies were a beautiful mix of blues, whites, reds, yellows, blacks, and grays, and provided a backdrop to the orange-gold sun before it sank behind the horizon. Peace descended. Maybe tomorrow would be a lucky baboon day.

CHAPTER 21

Visitors at Budongo

WHILE WE WERE AT BUDONGO, a Canadian named Pat Martin arrived and set up camp. He was far more organized and efficient than we were and had brought plentiful supplies and a lot of gear. Pat was studying birds and was interested in their conservation, but he also netted, killed and preserved specimens for his museum. One evening I watched with reluctant fascination as he whistled under his breath and concentrated on his task of skinning a small songbird. Suddenly he looked up and caught me staring intently. "Pretty little things aren't they?" he commented as he slit one open.

I wasn't sure what to think. It seemed a bit gruesome to me, but I suppose any collector kills and preserves specimens. Bats, butterflies, birds, flowers, bees, moths, beetles, are all killed, pickled, pinned, and put in cases or, with flowers, pressed between pages to dry.

On the second day of his stay, Mr. Martin, as I often called him, arrived back in camp carrying a guinea fowl he had shot. He presented it to me, telling me, "Get to work—skin it, joint it, and cook it."

I didn't like to be ordered around and, feeling short-tempered from the heat and extra weight I was carrying, glared at him. Cliff stepped in and made a joke to dispel the tension. "You have to understand, the nearest thing that Jen came to skinning a bird was when she took one out of a polythene bag in the supermarket!"

The Canadian laughed and, to his credit, stopped to explain what I needed to do, but I was furious when he stood there and waited for me to get to work. Yet I was not ready to admit defeat, too proud to say I couldn't handle the task, and too polite to tell him to get lost. But why did *I* have to deal with the guinea fowl and not *he?* Full of resentment and feeling sorry for myself, I knelt on the ground, bent over my distended body, plucked the feathers off the bird, slit it over the breast and stomach with a sharp knife, and took off the skin. After this, I jointed it. The worst part was taking out the glistening, slimy, warm, stinking guts. I disposed of them on the edge of the clearing, to be picked up later by scavengers. To my surprise, the whole operation had been much easier than I expected. I finished in ten minutes, covered everything immediately before the flies descended, and when the fire was ready, cooked the fowl. We had a delicious meal, and the fresh meat was a welcome treat after eating meat out of cans.

The Game Department Rangers preserved buffalo and elephant meat for later consumption by drying or smoking it. Cliff thought this was a great idea and tried to follow their example by searing meat and hanging it from a tree to dry. But he must have done something wrong, for the meat started to rot, became covered in flies, and attracted the vultures. We burned it on the fire.

Once or twice during our stay, we went to the sawmills to get the latest political news from Kampala. Everything seemed very uncertain. Mr. Knight said all the ministers who wanted a judicial inquiry into the gold scandal that Rikki had told us about before we set off had been jailed. According to the rumors, three had been killed. Knight said he anticipated rioting in Kampala and thought we would be safer to stay put until things died down. This was alarming news, but he was right; we could do nothing but lay low, keep abreast of what was happening, and make our moves accordingly.

After we left Mr. Knight, we visited an Indian who owned a paw-paw plantation. He extracted an enzyme from the fruit to use in baby foods in order to break down and soften meats. He gave us a sackful of the delicious looking paw-paw, or papaya, to help bait the baboon traps. Back at the campsite, Cliff put more poles around his trap to make it firmer and baited it with the fresh fruit. Some baboons watched from a distance but did nothing to acquire the paw-paw, though they approached the cage and walked around and inspected it. But baboons are not stupid. They were suspicious and probably wondered why this strange white man kept lurking around and putting food into a large wire box. He seemed up to no good, and they wanted no part of it. The upshot was there were plenty of baboons, but all efforts to catch them failed.

On the fifth day of our stay, a white man dressed in khaki shorts and shirt came to the camp. I estimated he was in his mid-twenties and was distinguished by his white-blond hair. "Hi, Cliff!" he said. "Babooning again, I see."

"Hello, John. Yes, but not much luck, I'm afraid."

The man looked at Caroline and me and raised his eyebrows. Cliff introduced us. "Jen, I want you to meet John Bindernagel."

Finally, I could put a face to the name. Caroline stared up at him quizzically but said nothing.

"Any news from Kampala? Any rioting?" Cliff asked.

John shook his head. "Nothing at the moment. It's remarkably quiet."

We were relieved to hear this, because we were planning to go back in two days, but I wondered whether it might be the calm before the storm.

John turned back to Cliff. "I saw baboons around your trap last week."

"Did they go in?"

"Afraid not. Not while I was watching, anyway."

"They just won't go in, and I'm not sure why. This time I've given them fresh paw-paw, but they just sit and watch."

"Sorry, I can't help. Seems they either don't like paw-paw or they're not hungry enough."

John had buffalo to attend to. He left after a few more cheerful words, and we promised to visit him at his home in Masindi.

A day before our departure, Pat Martin left, but before doing so he was visited by John Johnson, the head of the Forestry Department, who knew Cliff. He came over and said, "That friend of yours, Vernon Reynolds, wrote an article for the Fauna Preservation Society. Did you hear what happened?"

Vernon and his wife Frankie had studied chimps in Budongo, and we knew them well, having lived in the same house, first in Chalk Farm and then in Camden Town, London. Cliff said he knew nothing about the article. "Well," said Johnson, "since he wrote it, I've had lots of letters from old ladies who have accused me of poisoning the fig trees and depriving the chimps of their food."

"Really?" Cliff was genuinely surprised but, not knowing what Vernon had written, was unable to comment further.

"Tell your friend to lay off it!" Johnson said as he departed although he seemed more amused than worried or annoyed.

Cliff laughed. He knew he could do nothing to stop Vernon, whom he knew well from student days when they shared a dingy room in a dilapidated town house, later condemned, in Camden Town. The landlady, known as Mrs. Tink, worked at the Carreras cigarette factory near Mornington Crescent, and smoked heavily though she claimed to have no lungs. Her daughter and her somewhat intimidating Jamaican husband occupied the ground floor. An archeologist friend, Peter Ucko, who was given to wild exaggerations, claimed he had knocked on the door when visiting and it had been opened by the man wielding a huge knife. Knowing Vernon's sense of humor, we thought he would probably be pleased to hear he had stirred up mischief among old ladies.

Finally, it was time to pack up and go back. We had caught no baboons and left feeling disheartened. To lift our spirits, we stopped at the Masindi Hotel for a meal. We were dirty and battered. To be able to wash in clear, clean water was a great treat. We tried to make our clothes more presentable before sitting on the verandah in the comfort of wicker armchairs while we waited for lunch to be served and, for the first time in days, ate in a civilized fashion. From the verandah we had a good view of what was going on outside. A bus drove up and disgorged a crowd of American tourists talking excitedly to one another. With their scrubbed faces, clean pressed safari clothes, and their assortment of expensive-looking cameras and binoculars around their necks they could have landed from another planet. We were also surprised to see them when there was so much talk of political unrest and danger in the country. If tourists were still arriving we wondered if the situation was as serious as we had been led to believe. Surely they would not have come if there was a real danger of war?

The Americans were intent on taking pictures and spotted an African woman in a red turban, a white short-sleeved cotton top, and a long flowing skirt of reds and yellows sitting under a tree with her child. The Americans rushed over, clicking away with their cameras and indicating she should smile by baring their teeth.

"Gee, that's a real touch of local color," shouted one enthusiast as he headed off to look for more.

Considering ourselves seasoned bush people after our week-long experience, we felt rather smug and were amused by this behavior.

Many of the tourists wore the Bermuda shorts of which Americans were so fond. They clung to the legs, extended down to the knees and were invariably in plain loud-colored cotton or bright tartans. No self-respecting European would have been seen dead in them. Americans, on the other hand, seemed to consider the Brit's khaki shorts with their shorter and wider legs indecent. Our American neighbor made us laugh by commenting, "Wowee! Here comes Cliff in his hot pants!"

Anything less like sexy hot pants I could not imagine. But even if they held sex-appeal for humans, they certainly were doing nothing to attract baboons. In fact nothing seemed do that. We needed to find out why.

CHAPTER 22

The Elusive Baboon

ONE EVENING IN EARLY MARCH, Cliff developed a high fever. He complained that his limbs ached, his head ached, and he felt very cold; yet his skin was hot to the touch. He was no better the following day, but stubbornly refused to see a doctor. But late that afternoon he started to frown, kept looking at his hand, showed me a cut on it, and said he had been handling some monkey blood. Then he threw in the heart-stopper. People knew that monkeys could transmit viruses through bites or cuts, and there was a suspicion that some might carry herpes virus B. The virus had no obvious effect on the monkeys but it was usually fatal to humans, and there was no cure—you either got better or you didn't.

Suddenly I realized he thought he might have the virus. Until then I had no idea the situation was potentially so grave and felt my face go white from shock. Above all I knew we could waste no more time.

Why he hadn't told me earlier, when I could easily have obtained help, was beyond my comprehension. For some reason he seemed to deny anything was wrong until a crisis struck. Evening was upon us, but I daren't wait any longer. Feeling desperate and scared, I grabbed Caroline and, pulling her behind me, set off to the home of Dr. Patel, a medical doctor who lived on the estate. Fortunately he was there—but resting after a busy day when I arrived. He was not overjoyed to see me, but I assured him I would not have bothered him if this was not very serious and begged him to come. I was shaking, close to tears, and almost hysterical as I clung to my small child and insisted he must do something to help because I had nowhere else to turn. Seeing my agitation and noting how pregnant I was, Patel took pity on me and agreed to come to the house.

I led him into the bedroom where Cliff lay. Dr. Patel took one look and immediately grabbed his medical bag, took out some antibiotic, and injected a heavy dose. He told me this might be a bacterial infection, and that, if he was right, the fever should come down by the next day. If not, I was to find him at once—but, he added ominously, he wasn't sure anything else could be done. I was nearly seven months pregnant and had a two-year-old to take care of: Dr. Patel's words struck terror in me. What if Cliff failed to pull through?

After Dr. Patel left, I put Caroline to bed and went back to Cliff. His hair was plastered to his head, and he was starting to become delirious. Throughout that long night, I put cold washcloths on his burning forehead in an effort to bring down the fever that was raging through his body. His breathing was irregular, and he was extremely restless. If he heard me when I spoke, he showed no signs of it but lay either with his eyes closed or staring straight ahead, apparently unaware of anything around him. As the night progressed, he started to have hallucinations in which he lived in a private hell full of demons that caused him to pour forth gibberish and flail his arms. This was so unlike his usual, strictly logical self that my fears increased. What if he never recovered? What if his brain was permanently impaired? What if he went mad like the Victorian painter, Richard Dadd, who had undergone a dramatic personality change when traveling up the Nile and became increasingly delusional and violent?

THE ELUSIVE BABOON

While Caroline slept quietly in the other room, I was left alone with my fears and the sounds of shallow rasping breath punctuated by meaningless shouts. I had never seen anything like that, had no idea what the outcome would be, and was terrified. All I could do was persist helplessly with the wet cloths.

And then, just before dawn, at about five-thirty in the morning, there was a definite turning point. Cliff became less restless, his breathing became more regular, his temperature started to come down, and he slept. Emotionally and physically exhausted, I took the cloths away. When Cliff woke, he had a very sore throat and his tonsils were completely covered in thick white pus, like cream cheese. It looked as if thousands of white blood corpuscles had rushed there in his defense. I had never seen anything like it. After that, his temperature fell below normal and stayed there for about seven days, during which time he was extremely weak.

Later I heard that, when people got wind of what was happening from our neighbors, a rumor spread in the anatomy department at Makerere that Cliff had herpes virus B and was expected to be "a goner." People were beginning to realize that handling monkeys could be extremely dangerous, and their awareness had increased dramatically with the sudden and untimely death of Ron Hall, the chair of Psychology at Bristol University.

Ron had been only forty-seven when he died in 1965 just before we left for Uganda. His death had resulted from a fatal illness that followed a bite from an African monkey when it escaped from his laboratory and he tried to recapture it. No one was certain what the virus was, but Ron Hall's death sent shock waves through the community of psychologists, anthropologists, zoologists, and primatologists, and was still fresh in people's minds. When word got out that Cliff had been handling monkey blood, had a scratch on his hand, and was now extremely ill, everyone assumed the worst, convinced he was not going to survive.

These also were the days before HIV-Aids, when people were much less aware of the dangers of dealing with infected blood and the importance of wearing latex gloves. Even before Aids came on the scene, Ron's death drew attention to the fact that fluids from monkeys could be extremely hazardous, and attitudes towards dealing with monkey blood changed from cavalier to very careful. We never discovered what illness afflicted Cliff, but I came to believe that Dr. Patel's diagnosis of a bacterial infection was correct, and that he saved Cliff's life. What I *did* know was that Cliff was so shaken by what happened to him that, afterwards, he was always extremely careful when handling blood.

Meanwhile he had planned to revisit the West Nile District, leaving on March 20. When the time came, he was still very pale and weak and should have stayed home and rested, but he was determined to go. His illness had cost him a week when he planned to visit Budongo, and he was concerned that our year was passing quickly without capturing baboons and drawing blood. I was going into my third trimester of pregnancy, but decided Caroline and I would go with him. I could relieve him of the cooking and maybe some of the driving. If I drove, I could steady myself with the steering wheel. As a passenger, I had devised a system to help cushion my body from the bumps on the dirt roads by leaning back against the seat, putting my feet on the dash-board and bracing my body. I must have been in a state of denial or engaging in magical thinking, because I never allowed myself to think I would come to any harm. Anyway, I had been to the West Nile region before and knew it was nothing like the camp at Budongo. Cliff said I should remain at Katalemwa, but I too was stubborn. Rather than stay at home worrying, I would go.

On March 19, we assembled our gear, ready to leave early the following day. That night I woke to go to the bathroom, but when I stood up the house began to shake. Everything seemed to be in motion, with nothing stable to hold onto. The tremor lasted seconds, but the sense of disorientation and lack of physical control etched itself in my mind, though it had happened so quickly I began to think I may have imagined everything.

THE ELUSIVE BABOON

Over breakfast I told Cliff about my strange experience, but he had been asleep and felt nothing. However, when he went to pack the Rover, a neighbor hailed him and asked whether he had heard about the massive earthquake, around 4.45, that morning in the Toro district, about two hundred miles away and just southwest of Lake Albert in the Rift Valley. About 150 people had been killed, and there had been serious damage to dwellings, churches, and schools. Cracks in buildings were reported along the whole of the western border of Uganda, which followed the geologically unstable Western Rift Valley. I couldn't imagine the violence and terror felt by those near the center, if I had felt the tremor from so far away.

That morning we set off in a somber mood, thinking about the earthquake and feeling unwell in ourselves. This time we took a different route. Instead of going through Murchison Park, we skirted it and took the Gulu Road to the spectacular Karuma Falls on the Victoria Nile. These falls cascaded down and sent up amazing curtains of white foam caused by large stones that broke the water and were said to have been positioned in it by a great spirit, Karuma. We crossed by the Karuma Bridge, constructed only three years before, in 1963, and took the road along the side of Murchison Park. Again we saw many animals and at one point had to stop for elephants. Africans on foot, looking fragile in their presence, threw stones to get them to move. When they eventually shifted, we drove on until we reached the river and again took the ferry to Pakwach. This time we stopped for the night and set up camp in the grounds of the police station. We were erecting the tent when a raggedly dressed African with bloodshot eyes came over and shrieked at us in a voice that kept breaking into a high-pitched squeak. He hurled English phrases in our direction: "Jesus Christ," "Thirty shillings," "You think the African has no religion," and "Stupid, stupid, stupid." We had never before in Uganda encountered anyone so deranged. Disconcerted by the rant, we tried to ignore him, thinking he would go away, but he continued to shout at us.

"What do you think he wants, Cliff?"

"No idea. Maybe he's some sort of religious maniac, with all that talk about Christ." Cliff didn't hold with religion at the best of times.

"I just wish he'd go away."

"Maybe he's been drinking and wants money for more *waragi*."

"Seems to me he's had enough to drink already!"

The man continued to rant until we finally went into the tent and pulled up the zipper. After a few more bellows there was silence. When we emerged, he had gone, leaving us with no explanation for his strange behavior.

Early the next morning we took the Nile Road. We saw people in the fields picking cotton, which they formed into enormous bundles before covering them in plain or checked cotton cloths in a variety of reds, greens, blues, and yellows. Gangs of women with hips swaying walked along the road. On her head each woman carried one of these large bundles of cotton, which she was transporting either to market or to a center for processing. The floppy bundle molded itself to her head and came to rest on her brow in front, and on top of her ears at the side. It looked like an oversized puffy hat more suited to a cold climate than the intense heat of the West Nile.

We reached Arua early and decided to have a meal and visit the open-air market to fill in time before we met Mr. Nkalubo later that day.

To our surprise, the market contained very little. Men were selling fish, and women sat on the ground with small piles of food laid out in rows—a few peanuts, four of five tomatoes, or a few beans—but there was no activity comparable to that in the Kampala market. Eventually we went to find our friend Mr. Nkalubo, who said he had been married since we last visited and was sorry we had not gone to his wedding. Thinking we would like to see a real African wedding, he had sent us an invitation. It must have gotten lost or been waylaid, because we'd never received it. For me, this was one of the biggest disappointments of our stay in Uganda.

Mr. Nkalubo said he had negotiated with a band of fifteen men to catch baboons. Finally, all seemed set for success. While I stayed in camp with Car-

THE ELUSIVE BABOON

oline, the two men set off on the great baboon-trapping expedition and reported back in the evening.

On the first day, they met up at the appointed place with the hunters, who were carrying nets, bows, and arrows. The hunters referred to Cliff as Bwana Kubwa (the big boss), which seemed to be a remnant of colonialism. The plan was to use hunting nets, generally used to poach small antelopes. These nets were, strictly speaking, illegal, but Nkalubo had persuaded the men to bring them out with the promise that he would turn a blind eye. The hunters were to split into two groups. One group would attempt to drive a troop of baboons into the nets, which were to be set in a long line. The other group of hunters would then catch the entangled baboons.

This plan might well have worked for antelope, but it didn't work for baboons. Only a few nets were used and, not surprisingly, most of the baboons simply detoured around or climbed over them to escape the hunters. Mr. Nkalubo used his rifle to kill a large male baboon that had been cornered up a tree. The next day he shot another, so we ended up with two blood samples, which was better than none but Mr. Nkalubo was disappointed at the failure of the grand plan. He would have been willing to shoot more baboons, as they were classed as "vermin" by the Game Department. But Cliff had no wish to kill more animals simply to obtain samples.

To begin with, we were baffled by the way in which the men behaved but, when we got to know more about local customs, thought we might have some answers. In retrospect, it seemed that the hunters had an ambivalent attitude toward baboons. They wanted to get rid of them because they caused extensive damage to their crops, but they also feared them because they were said to have magical properties. Cliff personally witnessed the Africans' reluctance to touch the dead baboons. When he decided to bring one back to Kampala, he had to pay two men additional money to carry it and put it on top of the Land Rover, and when he took blood from an animal, the local people appeared to look on him as some kind of wizard and steered clear. This reputation was reinforced when he found a tortoise that he put in a bag to give to Caroline, causing laughter from the men. Then he found a python's head, which he picked up and which evoked nervous laughter from the watching crowd. It turned out that parts of all three were used in medicine and thought to have magical properties.

This attitude towards baboons reminded me of how the ancient Egyptians revered baboons and kept them as pets and as sacred animals in temples where they were worshiped. The Egyptians often associated them with sexual potency and prowess and, in addition, with Thoth, the god of wisdom, science, and measurement. Thoth was a moon god, and his baboons often wore crescent moons on their heads. But some believed baboons were solar animals probably due to the way they sat and warmed themselves, facing the morning sun. There was a lot more to baboons than I had realized.

We had more evidence of the perceived value of baboons when we set off back to Kampala early the next morning with the baboon attached to the roof of the Rover. It was a magnificent male that caused quite a stir and drew crowds along the way. Someone tried to buy it on the Pakwach ferry, because they said the liver was especially good for treating illness. I had already heard one of the game guards say, "It is better than Aspro for pain, madam."

Unfortunately the animal deteriorated rapidly in the damp and heat. Although Cliff had injected it with a good deal of preservative and it had been dead for only a day, by the time we reached Kampala it stank horribly. People stared in fascination but also turned up their noses in disgust as we drove past. Cliff took it to Mulago to be made into a skeleton.

In spite of all the good will, our trips to the West Nile were not productive. However, we were learning some lessons about how to capture these elusive baboons. Nets obviously didn't work, especially if you wanted to catch a lot of animals. More time was needed to set up a site with traps and get the animals used to them. In addition, an organized team committed to trapping was important. Given the mixed feelings of the Africans, we concluded they might

THE ELUSIVE BABOON

not be the best people for this. Our visits to Budongo made us realize as well that, if the animals had plenty of food, they were hard to lure into traps.

Thus Uganda was a good place to observe baboons but not the best place to trap them. In London we had anticipated none of this. But all was not lost, because Cliff had been gathering blood from vervet monkeys at the Virus Research Institute in Entebbe, and from his trapping expeditions with Jimmy.

Meanwhile serious trouble was beginning to brew as the political environment grew increasingly unstable.

JENNIFER JOLLY

CHAPTER 23

A Recipe for Trouble

WHEN WE ARRIVED IN UGANDA, the country had seemed calm. During the first few weeks I had gotten to know people, learned to shop in Kampala, arranged for local help, taken care of Caroline, and cooked, all while trying to deal with violent vomiting in the early stages of pregnancy. I had assumed a second pregnancy would be easier. Mine was not. On the advice of friends, I had chosen to go for ante-natal checkups and for the delivery at Mengo Hospital, an old missionary hospital with excellent maternity facilities. It was situated close to the Anglican cathedral and next to the palace of the Kabaka, or King of Buganda, on Mengo Hill, one of the seven on which Kampala, like Rome, was built. Any thoughts of political unrest had been far from my mind.

However, around the beginning of December, stories began to circulate that Uganda was threatening to leave the British Commonwealth if Britain's Prime Minister, Harold Wilson, failed to take action against Ian Smith's government in Rhodesia (now Zimbabwe). Smith's white-minority government had rejected a proposal for greater African representation in the colony's parliament that would have meant black majority rule in the self-governing British colony. Wilson and Smith met in London to try to resolve the issue, and when these negotiations broke down, Prime Minister Smith returned to Rhodesia and declared its independence from Britain.

When Uganda threatened to leave the Commonwealth, consternation broke out among the British in our community. If Uganda carried out its threat, we heard we would all be expelled from the country. Rumors coming from Kampala indicated things were deteriorating rapidly, and we were advised to get ready to pack. Tension increased as people tried to obtain information, waited to hear the worst, and huddled in groups trying to decide what to do. Neighbors told us a car bearing G.B. license plates had been set on fire outside the British Embassy in Kampala, where it ended up a blackened, burned-out shell. More rumors followed.

In spite of their fears, people tried to stay calm, exchange information, and reassured one another that all would be fine. The stoicism of the British to grin and bear it, together with their inclination to do little until their backs are against the wall, served everyone well. I stuck my head in the sand, refusing to believe the Ugandans would want the British to leave when so many provided help in agriculture, industry, medicine, and teaching, but I was very worried. Our life was beginning to fall into place, and I had no idea what we would do if we had to go back to England with nowhere to live.

In the end, the incident turned out to be a tempest in a teacup. Nothing more happened. People returned to their daily activities, and the panic died down. Nevertheless, I began to realize things might not be as calm as they appeared on the surface, and my unease increased towards the end of the year, when stories began to surface that a military coup was going to occur. I started to read more, listened carefully to what others said, asked questions, and paid more attention to my surroundings.

One morning, I was sitting on the sofa, waiting for my heaving stomach to calm down after the usual sickness, and considering why there might be some truth to the rumors. From what I had gathered, the key factors contributing to the likelihood of political upheavals were the history of the country and the changes brought at independence.

Uganda as it existed in 1965 consisted of two main regions, differing deeply in culture, language, and traditions of government. The north was a patchwork of different ethnic groups, mostly speaking languages of the nilotic family, with no tradition of strong, central government. The south was a region of Bantu-speaking tribal kingdoms that had developed over centuries of migration, conquest and conflict. The four main ones were Toro, Ankole, Bunyoro, and Buganda, the largest. The country first came to the notice of Europeans in 1862, when John Speke, an English explorer looking for the source of the White Nile, entered from the South via Tanganyika and reached the kingdom of Buganda, which was dominant at that time, and met the King, Mutesa I. They got along well, and Speke stayed with him, exchanging gifts, receiving a wife, and taking part in ceremonies. After that, he continued on his way and finally proved the White Nile flows out of Lake Victoria. The hotel where we first stayed was named after him.

Thirteen years later, in 1875, the Welsh-born American explorer Henry Morton Stanley reached Uganda. He too met Mutesa and, impressed by the king's interest in the Christian faith, wrote to the *Daily Telegraph* in London to urge that missionaries be sent out there. His challenge was soon met. When we arrived, the influence of the Catholic and Anglican missionaries was still very evident in the schools and hospitals. Traders and prospectors were also attracted to the country, and, in 1894, Buganda fell under British control as a protectorate. Two years later, this authority extended to cover most of Uganda. But under the British colonial system of indirect rule, the King or Kabaka of Buganda, whose people were called the Baganda, remained a significant figure.

In 1962, exactly a hundred years after Speke arrived, and two years after Macmillan's "Winds of Change" speech, Uganda was granted its independence. A significant change in governance then took place. A federal republic was established in which the current king, the Kabaka of Buganda, Mutesa II was named as its first President in 1963. Milton Obote, a Langi from the north of the country, was appointed prime minister. The other three kingdoms, Toro, Ankole, and Bunyoro, were smaller than Buganda, but each had a king, and through them the country was united under the central government. This was the state of affairs when we arrived.

I assumed that an attempt had been made to bring north and south together by giving key government roles to a president from the south and a prime minister from the north, but in reality they formed strong opposing factions: The northerners and southerners disliked and distrusted one another. The Baganda who lived in the area around us especially distrusted and disliked the Acholi and Lango from the north of Uganda and regarded their traditionally uncentralized, village-based societies as hopelessly uncivilized. These Bantu-speaking people especially disliked the Acholi and Lango in the north of Uganda, whose language and traditional culture were totally different from their own. Thus it wasn't surprising that the Baganda felt antagonism towards Obote, who was from the northern Lango. The situation was further complicated because Obote was not only the prime minister but also head of the Ugandan army, of which the majority of members were northerners. The practice of recruiting army and police chiefly from the northern tribes had begun under British rule, as a counterbalance to the southerners, especially the Baganda, whose traditions of western education and centralized government naturally fitted them to enter the bureaucracies of the judiciary and civil service. This divide-and-rule policy worked reasonably well under the protectorate, but with British authority withdrawn it seemed potentially a recipe for disaster.

In contrast to their intense dislike of Obote, the Baganda revered the Kabaka. Not only was he their political leader, he was also a sacred king, the embodiment of his country. We witnessed first-hand the deep emotional loyalty of the Baganda to the king when a public holiday was announced on November

19 in honor of the Kabaka's birthday. Wild parties, accompanied by singing and dancing, took place all day as the Baganda in the surrounding villages celebrated the occasion. Drums were beaten, and as the noise increased in intensity it echoed across the lush green slopes of the valley in the distance and spread into the surrounding areas. This constant rhythmic drumming was accompanied by shouting, hollering, and whooping that lasted into the night.

Although they could shift rapidly from laughter and joy to anger and reprisal, people generally loved to have a good time. Their laughter, accompanied by singing and the sounds of voices lubricated with home-made beer and *waragi*, the local banana liquor, prevailed on the Kabaka's birthday. All the Baganda nearby stressed to me, "It is a very special day madam," and I believed them. I had never experienced anything like it. Their excitement was contagious, and when the vibration from drum beats pounded through the warm tropical air, I sat tapping my feet to the rhythm. The noise finally died down in the middle of the night, leaving behind a comfortable sense of calm.

After this, I felt convinced the Baganda would fight to the death to protect their Kabaka, and that, if Obote ever tried to usurp his power, we had reason to believe that violence would rapidly erupt. We knew that people often took the law into their own hands, and violence occurred on a daily basis. If a man was caught stealing, others thought nothing of beating him severely and even killing him. A few weeks after we arrived, one of my neighbors told me a story about a lorry that had hit a car. After the crash, the lorry driver jumped out and ran. A crowd chased him, caught him, and axed to him death. Based on this, it was likely that any attack on their revered Kabaka would rouse the Baganda to extreme and widespread violence.

Getting up from the sofa for a glass of water and a dry cracker to try to calm my stomach, I noted I was still thin. I found it hard to believe I was going to get much larger over the next few months. If only I didn't feel so sick. Absorbed in thoughts about the antagonism between north and south, the great

THE ELUSIVE BABOON

loyalty of the Bagandans to the Kabaka, and the rapidity with which violence could erupt, I wandered through to the kitchen. John was standing there with a smirk on his face, holding a dirty wet cloth that dripped a mix of red-colored soil and water onto the floor, as if to provoke me. I ignored him. My mind was focused on the deep-seated tensions in the country. They seemed to be bubbling under the surface, waiting to erupt. I wondered when something would happen to bring the two opposing forces of north and south into outward conflict, and what we should do to prepare for trouble.

CHAPTER 24

Checking the Rumors

After the Rhodesian scare, we took our first trip to Arua and returned to celebrate Christmas in the heat, joining our Welsh friends from the ship. They lived about twelve miles north of us at the agricultural station of Namulonge. It was strange to eat Christmas dinner, wearing paper hats and pulling crackers, with no sign of frost, snow, or Christmas trees. We ate well, relaxed by their pool, lazed in the sun, and went swimming while their African help served food and drink. With their good salaries and lots of assistance, I thought again of how the ex-pats led a comfortable, easy life.

In January, we traveled to Elgon, on the fossil-hunting expedition, without any trouble, yet rumors kept surfacing that a military coup was going to take place in Kampala. Each time they did, everyone went into a panic. Each time nothing happened and, like the villagers when the boy cried "wolf," we started to become skeptical and take no notice.

But at the beginning of February, while Cliff was away in Budongo, I was resting on the sofa and reading a story to Caroline about a busy mole named "Diggy." Diggy was about to seize his pick to make a hole in the ground when we were interrupted by the sound of footsteps on the verandah and Rikki's thin, bespectacled face, her dark hair pulled back in a bun, appeared unexpectedly at the French doors. She opened them, strode purposefully into the cool interior, announced she had some disturbing news, and sat down. Extracting a packet and some matches from the pocket of her khaki trousers, she shook a cigarette from the pack and lit up. After taking a deep drag and blowing out a plume of smoke from the side of her mouth, she said that rumors from Kampala indicated a military coup was imminent. People on our estate were stocking up with provisions, and food was flying off the shelves of the supermarket. One woman up the road had laid in enough supplies for six weeks, because roadblocks were expected to go up on the Gayaza Road, the only route from our estate to Kampala. If the rumors were true, we would be unable to get into town for food supplies. The estate would be isolated, and we would not be safe.

I felt as though I had been punched in the stomach. Caroline began to fret when she heard me gasp and stare open-mouthed at Rikki. I told her to be a good girl, take her book, and play with her toys; I turned back to Rikki and asked why they were putting up the roadblocks. She said they were trying to catch government ministers who had fled the capital once it was overthrown. She added the ministers already feared for their lives if they returned to their own homes. To protect them, some university people who lived on Makerere Hill had taken friends from the ministry into *their* homes to stay overnight. She gave no explanation of exactly what the military coup was about and why the ministers were fleeing.

Everything seemed confused, and I struggled to make sense of the information. Apparently, ministers were hurrying away from Kampala, and no one knew why. There was going to be a military coup, but no one knew how serious it was or exactly what it was about. The only thing we knew for sure was that past threats of a coup had led to panic and resulted in nothing. I thought there was a lot of supposition in it, based on rumor, and said I wanted more facts before getting too wound up, but I also tried to quell the panic rising in my gut. I was on my own with Caroline, had no vehicle if I needed one to escape because Cliff had taken it with him, and, like others, we had no telephone to contact people. I was almost six months pregnant as well; with no transport, I'd be unable to escape if the need arose. Anyway, where would I go? And why was this threat different from all of the others?

Rikki agreed things were not clear but said the situation seemed different. I had to admit that I couldn't remember hearing specific information before, like ministers having to abandon the city and be taken in for protection by others, rumors of road blocks, no food on the shelves, and disturbances in the capital. We decided to find out more from neighbors on the estate. Grabbing Caroline's hand, I followed Rikki into the hot sun, dragging my small child in tow and hauling my body along. I had swelled up at an alarming rate over the previous few weeks, going rapidly from thin to fat and from fast to slow. We came across some neighbors huddled in the shade of trees, discussing the situation. One said he had been into Kampala the previous day, and nothing had changed. A local drama king flung out his arms and declared in a deep baritone, "You could smell the tension in the air." But no one could tell us more than we already knew.

With conflicting information and no idea of what to believe, Rikki and I decided there was only one way to find out what was happening: We would go and see for ourselves. We found a neighbor to take care of her son and my daughter, and set off in her car along the road to Kampala. As usual, women were walking alongside the road with baskets of food on their heads, and men were cycling along with bundles of wood or *matoke* strapped to their backs. There were no roadblocks. We drove into the center of Kampala and headed to Cashco's Supermarket, expecting to see shelves left bare by panicking shoppers. The shelves were well stocked. Only a few were shopping. There were no outward signs of any of the struggles we had been warned about, and we saw no signs of people fleeing.

THE ELUSIVE BABOON

We concluded that the rumors were false, that there was nothing to worry about, and headed back to Katalemwa. It wasn't until I saw my small daughter playing happily with other children that the potential seriousness of what we had done hit me. Had anything happened to me, she could have been left alone. What if the rumors had been true and we had been caught, held in Kampala, or even worse? I was nearly six months pregnant. What about my unborn child? Perhaps we had taken a stupid risk, but Cliff, who returned soon after, had experienced no problems along the Gayaza road either. We concluded it was yet another false alarm.

But the rumors continued, and while we were planning our trip to Budongo towards the end of February, the scandal broke in which Obote and Amin were accused of selling gold for their own personal gain. In Budongo we had heard that the ministers who wanted a judicial inquiry into the gold scandal had been jailed, and that Obote was said to have killed some of them in retribution: a few managed to escape. Looking back, I realized this scandal was probably behind the rumors Rikki had heard when she reported that ministers feared for their lives. Again we expected trouble in Kampala, but once more things quieted down.

March went by, and we took our second trip to Arua after Cliff's virus B scare. By April 3, Obote was still in the country despite rumors he had fled with his ill-gotten gains from the gold scandal. We knew these rumors were untrue, because we saw him at Entebbe Airport when we drove there to send blood samples to London. He had flown in on a military Comet and been met by a large guard of honor and a big brass band. In spite of the oppressive heat, the men had been dressed in full khaki military regalia, including hats, buttoned-up jackets, and shiny black boots as they stood on tarmac runways that glistened and gave off a shimmering haze in the sun.

Then in mid-April, Obote unexpectedly seized the presidency from the Kabaka and declared himself President of Uganda. Things were beginning to heat up. It was unlikely that the Baganda, with their very strong loyalty to their king, would take it lying down. The Kabaka had his own police force, which though no match for the Uganda Army, was armed and could be trusted to be loyal. For the first time, we saw young men armed as soldiers in cars and wagons in Kampala. They waved guns around in a casual manner, giving the impression they were trigger happy, and raised our fears of getting caught inadvertently in a cross-fire. But there seemed little we could do apart from carry on with life as usual, keep our heads down, and stay alert. We were sure this would be a breaking point, but the outward tension died down.

On April 20, Cliff set off for Budongo again and returned five days later. He had been back only a short time when we heard dreadful news. John Marshall, an American professor who had been setting up an electron microscope at the medical school and who was one of our immediate neighbors, had been killed while on holiday in the Kigesi district. He had been driving a Volkswagen mini-bus that rolled over into a ditch and pinned him underneath. By the time the rescue crew came, it was too late. His wife and five of his six children were with him and escaped with minor injuries. Their eldest son had just left for college in the States.

I couldn't begin to grasp the horror and devastation caused to this close-knit family. They were dependent on him for their livelihood, and he had spent quite lavishly and generously.

On one occasion, John had arranged for musicians and dancers to perform in his garden one evening and we were invited over. The head drummer had been the Kabaka's chief drummer, and some of the dancers had performed at the Commonwealth Arts Festival. Colored lights had been rigged up in the trees, and a large fire set up at one end of the garden where children cooked sausages. Men in white robes played drums. Women with lithe bodies, clad in sleeveless white tops and long striped skirts of bright red, black, yellow, and white shimmied to the rapid beat of the drums and flapped their hands over their mouths to make loud whooping noises. The performance was full of energy and excitement. People in the audience cheered, threw money, and begged for encores. Everyone had a wonderful time.

Now the unthinkable had happened to this popular, still-young man, and people were overcome with grief. His family was left with no option but to return to the United States, where they would be dependent for a while on his widow's parents.

This event marked a definite turning point in my own life. I realized that, as soon as possible, I needed to make sure I was self-sufficient if faced with a similar tragedy, and my mother's warning came back to me that it was important to have a career because "you never know what might happen."

So far we had had no trouble on the estate, but fresh rumors began to circulate about people being robbed on the road or having their cars tampered with. Events were taking their toll. I became more nervous and was afraid of staying alone when Cliff went off on his expeditions. My mind began to be crowded with morbid images. Cliff had already had two serious illnesses, one of which nearly cost him his life. I dwelled on the death of our neighbor; I thought of people killed on the roads, people killed for theft, people being beaten up and robbed.

Maybe events weighed more heavily on me because I was tired from carrying extra weight and worried about giving birth in the midst of so much turmoil and uncertainty. We were seven miles from Mengo hospital where I was to have the baby, and Mengo was near the Kabaka's palace, the Lubiri. Given the antagonism between the Kabaka and Obote, this was likely to be the center of any trouble. For the first time in my life, I became seriously aware of my own mortality and that of people close to me. I waited on tenterhooks for the next shoe to drop.

CHAPTER 25

Fighting Breaks Out

WHEN DISASTER FINALLY STRUCK, we were caught unawares. Our lives continued to be busy. Cliff took blood samples from vervet monkeys; Caroline went to school and to parties, and played with friends; and we visited or were visited by others for lunch, dinner, coffee, and tea. We had had a number of political scares, but nothing had come of them.

By May it was too risky for me to travel far, but Cliff drove to Rubirizi in the extreme southwest of the country, the third area we had selected for baboon trapping. On a previous visit he had left word that he would return to buy skulls of any baboons killed in pest control operations, and if possible to take part in a baboon hunt. People wanted to help and were generous with what they had. When he stayed in a small village beyond Mbarara, his hosts gave him chicken and fruit on his first day and eggs and more fruit on the second. Some pest-control people had been hunting and killing baboons. They rough-cleaned some skulls, charged Cliff a shilling each, and he returned with several of these ghoulish-looking objects rattling around in the back of the

THE ELUSIVE BABOON

Land Rover.

The hunters said they would take him to the baboons and walked him for miles through rough terrain marked by small hills. They led him along the edge of a ravine where he had to hang onto overhanging branches to avoid plunging into the roiling waters below. They tramped over vast areas covered in head-high elephant grass, where the leader cleared a path by putting out his hands, falling flat on his front to flatten the grass, and then proceeding like a seal on dry land to clear a pathway. But they found no baboons. Well-meaning locals would point and say, "We heard them over there." But at the spot indicated, another group would say, "We heard them over there," and point back in the direction from which they just came. One man said he had heard baboons, but they turned out to be colobus monkeys. Once again, baboon hunting was a bust.

On May 11, soon after Cliff returned from this expedition, panic broke out on our estate after several burglaries occurred, and a meeting was called. One woman declared in a loud upper-class English accent that the wheels had been stolen from her car. "We must *do* something!" she barked, sounding like the imperious Lady Bracknell in *The Importance of Being Earnest*.

"But what?" people asked. The general consensus was that we had to be more vigilant. Cliff told me he would outfox anyone who tried to meddle with our Land Rover. Taking string, paper clips, rubber bands, and a mouse trap, he created a complex alarm system, saying that, if anyone tried to break into our garage, the trap would spring, set off the hooter, and scare the intruder away.

Many of his projects depended on rubber bands, paper clips, string, bits of wood, chicken wire, and a large hammer. A hammer was invaluable for whacking things, sometimes causing wide-spread destruction, as it had the time he hit a water pipe while enthusiastically demolishing a wall. I'd grown up with a down-to-earth, practical father whose strong, capable hands meticulously carved wooden toys and carefully repaired complex machinery.

All of this larking around with rubber bands, bits of string, and so on baffled me. I was skeptical. "You don't really think it will work, do you? Doesn't seem all that practical to me."

He ignored me and continued. Apparently I didn't have his creative imagination. Unfortunately, as he was putting together his Rube Goldberg contraption, he cracked the hooter and defeated its purpose. "I suppose we don't *really* need a hooter anyway," he said. I was speechless. He reminded me of Vernon creating a death trap when he blithely proclaimed, after the brakes on his MG broke, that it didn't matter, cars were meant for going and not for stopping. Imprisoned in my oversized body, there was nothing I could do but shake my head, mutter, "I don't believe this," and trundle back into the house.

Meanwhile, I continued to visit the doctor, and in the mornings Caroline went to a nursery school in Kampala run by a lovely, motherly English woman, Mrs. Saldana, who was married to an Indian high court judge. This freed up time for me to pursue some work of my own on the relationship between personality and perception. I was interested in the physiological and neurological basis of behavior, but the technology for studying the brain, and the ways in which it received and responded to different stimuli, was limited at that time. Specifically, I was curious to know how anxiety and stress influence what we see. I tested this by setting up a device at Makerere to measure the actual size projected on the retina by an object set at various distances from the person being tested, and the size the person reported it to be. I also administered written personality tests. Nearly all the medical students who participated were of Indian ancestry, either Muslim or Hindu. I had found Muslims scored higher than Hindus on a measure of extraversion. Although some people claimed introversion was associated with neuroticism, I didn't find this. My results showed that, regardless of whether they were introverts or extraverts, people who had higher scores on anxiety and neuroticism reported the object was smaller than those with lower scores on these measures. I thought the preliminary results were interesting, because they raised questions around the old debate over how much culture and genetics influence our perceptions and behavior. I wondered what the implications might be. For instance, if the results held true in a larger population, did this suggest that some people are naturally more predisposed to anxiety, and that this affects their perception of the outside world? If so would they react differently to different stressors? Might they be more prone to psychological and medical disorders? What might be the implications for treatment? But I needed to do far more research.

Unfortunately, my work had to be abandoned after what happened next.

On Monday, May 23, a few days before I was due to give birth, we took Caroline to school early in the morning, bought groceries, and drove the seven miles back to Katalemwa, where Cliff was working at home. At midday, he set off to collect Caroline while I stayed to prepare lunch. We planned to eat at around one o'clock, and I expected them back around twelve forty-five. The meal was ready, but one o'clock came and went and there was no sign of the Land Rover. This was unusual, but I assumed our unreliable vehicle was giving problems. Then I realized it was *strangely* quiet. There were no cars going along the road of the estate, and no one was walking. Something was wrong. The only way I could contact anyone for information was to leave the house and, given the odd silence, I was afraid to do that. Time passed, and with my stomach in knots I helplessly paced the cool floor in bare feet as my anxiety turned to alarm. And then, just after one-thirty, I finally heard the welcome sound of the Land Rover coming slowly up the driveway. It drew to a halt. There was still no sign of any other cars. A minute later, Cliff appeared, carrying Caroline, who was crying. His hair was awry, he was covered in dust, his face was pale, and he appeared shaken as he handed Caroline to me, sat down, wiped his brow, and took a deep breath. I held Caroline, who began to calm down, and then he told me what had happened.

After he left me, Cliff had headed onto the Gayaza road to Kampala only to discover that roadblocks were being put up not far from our estate, and men were busy digging up the tarmac to stop the traffic from moving. Knowing that Caroline was in the middle of Kampala and cut off from us by seven miles, he had decided to get to her no matter what. He'd turned off the main road, taken a series of detours into the bush at the side, and, once past the blocks, turned back onto the road, on which he had driven as fast as he dared into Kampala.

In the interest of time, he hadn't stopped at the school to find out what was happening but gathered Caroline, put her into the Land Rover, and faced the problem of how to get back. Had I gone with him, we would all have stayed in Kampala until we knew it was safe to drive home, but he couldn't get in touch with me, so he had decided to take a risk and head for Katalemwa. He'd set off along the tarmac road and charged towards a man digging it up, who leaped for his life when the Land Rover bore down on him. Nearer to the estate, digging had hardly progressed due to a tremendous and unexpected thunderstorm, but he had still had to detour into the bush. While there he'd hit a rut. Caroline had been thrown forward against the windscreen and hurt her head because there were no seat belts. Fortunately, the vehicle had kept going and they had arrived shaken but relatively unscathed.

The meal sat on the table uneaten because no one felt hungry while we waited in a state of suspense, wondering what to do and what to expect.

And then the eerie silence outside was broken by the sound of vehicles and voices as people started to drift back to the estate, congregate, and exchange stories. One of our neighbors had had a brick thrown at his windscreen when he drove down the road only a short while after Cliff left Kampala. Several people had taken an hour and thirty minutes to complete the fifteen-minute journey home because they made their way back to the estate through the bush by traveling in convoy along a series of small tracks.

Our friends the Walkers had been trapped in Kampala in the morning and only managed to get back in the afternoon. They had sheltered in the Anatomy

THE ELUSIVE BABOON

Department at Makerere and heard a terrific noise coming from Mengo Hill, two hills away. It had sounded like the roar of a huge football crowd with drums beating furiously. They'd heard rumors that fighting had broken out near to the Kabaka's palace. This frightening news confirmed my suspicions that this was likely to be the center of fighting and very close to where I planned to give birth in Mengo hospital. I was horrified to think I could have been in the hospital when fighting erupted. Now I worried about what I would do if the fighting continued and I couldn't get to Mengo as planned.

Rumors flew. We heard many people had been taken to Mulago Hospital and that many others had been killed, but it was difficult to know what exactly was going on when official communications were so bad and we had to rely on hearsay. Later we learned the fighting had been precipitated by the arrest that morning of some of the Saza chiefs, who were district heads in the Kabaka's administration. People speculated the arrests had been made by Obote's police in retaliation to the Kabaka's appeal to the Secretary-General of the United Nations, U Thant, for support when Obote seized the presidency. Riots had broken out on the Entebbe Road near to the clock tower, a landmark in Kampala, when people heard of the arrests. In the midst of the confusion, one thing was clear. Events had taken a decided turn for the worse. We had been expecting trouble for months, but nothing had happened. Now it seemed as though everything was coming to a head with Obote intent on deposing the Kabaka and seizing absolute power.

On Tuesday, May 24, my twenty-seventh birthday, we woke to the usual brilliant sunshine, not realizing this was to become one of the most significant days in Uganda's history. It was the day when Obote ordered his army to attack the Kabaka's palace, the Lubiri, on Mengo Hill.

That day we heard the palace was under siege and on fire. Shouting broke out all around the estate. Fred, our *shamba* boy, came to tell us he had heard the Kabaka had been arrested and beaten. Conflicting stories about the Kabaka went around, and we were not sure which side the police, special forces, and military were on. At one point a great wailing went up from the Baganda in the vicinity when a rumor circulated that the Kabaka was dead. This proved not to be true, but during the battle many priceless historical artefacts were stolen or destroyed.

That morning, Cliff, Rikki, and a neighbor risked going into Kampala to buy food. They returned to say Kampala was quiet and deserted, but that they had seen a lorry load of men with guns at the ready, and a burned-out bus being towed into Kampala. Cars had been burned out, turned over, and abandoned. Numerous deaths and beatings were being reported. They'd heard that two European surveyors had been killed in a police station on the Jinja road the previous day when a gang attacked the station and killed everyone in there.

We heard that a bus going to the airport at Entebbe was overturned, and no one knew what had happened to the passengers. According to reports, there had been some Americans among them. A neighbor told us that some Europeans who gave the Baganda lifts in their car had been stopped by the police and beaten up because they were viewed as showing sympathy to the Baganda and taking the side of the Kabaka.

A rumor went around that Judge Sheridan, who was crippled, and was on the *Kenya Castle* on our trip to Mombasa, had had his car stopped and that he had been dragged out and beaten up. He was said to be recovering in Masaka Hospital, about eighty miles away. Caroline's teacher reported that two hitchhiking European teenagers had been killed. We heard that, in the storming of the Kabaka's palace, Obote's army had been led by Colonel Idi Amin. We knew little about Amin other than that he was rumored to be a collaborator with Obote in the gold scandal.

Early that evening we heard the alarming sound of exploding mortars. A curfew was imposed between seven in the evening and six in the morning. During those hours it would be extremely dangerous to leave the house.

By May 25 we still knew very little about what was going on and were told Uganda radio gave hardly any information. We heard the police were causing a lot of upset. Joe Mungai, a Kenyan professor in the Anatomy Department, reported that two men who tried to dig up the road between Wandageya roundabout and Makerere had been shot dead by the police and left to rot for two days.

Cliff said Kampala had been quiet when he visited in the morning, so we decided to take Caroline to a party to which she had been invited. We didn't stay long in view of the curfew and the strange unsettling silence in Kampala. The woman who held the party told us she worked with a man who had intended to drive to Entebbe the previous day. He had been stopped at some roadblocks on the outskirts of Kampala and fortunately got out of his car and threw himself behind it when he saw the police advancing. They riddled it with bullets as they went past. He survived and managed to get back. Another car was found on the Entebbe Road full of bullets and with all of its passengers dead. Getting to Entebbe Airport was difficult if not impossible. After we heard stories about people being shot on the roads, we realized we had been foolish to go to the party. On our way home we had passed security police with guns at the ready, but fortunately they'd left us alone.

No one seemed to know the whereabouts of either the Kabaka or Obote. Rumors said the Kabaka had fled the country. Some said Obote had gone to Gulu in the north, others that he was in Mbale, to the southwest. On May 26 Cliff and Rikki went into Kampala and said things seemed fairly quiet. While there they were told Obote had addressed Parliament the previous evening, so he must have been in Kampala. We heard that no more people were being allowed to leave Uganda for the present. The last plane had departed at midnight, so we were stuck and couldn't get out even if we wanted to. Meanwhile, the baby was due and Cliff's idea of getting me flown back to England with Caroline was no longer feasible.

We heard that all stations run by the Kabaka's police had been taken over and demolished. Three reporters had been deported. An eyewitness said he had seen lorries full of bodies coming from the Kabaka's palace on Mengo Hill. All were presumed dead and on their way to be burned or dumped in Lake Victoria because no one reached Mulago Hospital for treatment. Someone saw a German tourist taking a photo of security police on Kampala Road. The police had stopped him, beaten him up, put him in the back of their van, and he was not seen again.

The chief tsetse control officer, and his assistant, were in Mulago Hospital. The former was suffering from head injuries after being beaten up. His arms had been slashed with *pangas* and we heard he would probably lose one from below the elbow. I wrote in my diary:

I continue to panic about going to Mengo Hospital to have the baby. It is situated so close to the Kabaka's palace, which is the center of the fighting. Even if the hospital is left intact and fully staffed, it might not be safe. In any case, if I go into labor at night it will be risky to drive along the roads to get there during curfew. Even if we have police permits to travel, everyone thinks we would probably be shot before anyone bothers to look at them.

My world had turned upside down. What would I do if I couldn't get into Mengo? We decided to set up emergency plans, so that if the baby started to come at night, it could be delivered on the estate. Rikki confidently claimed she could do this, but I was sure she has never delivered a baby before and was afraid of unforeseen complications. I decided I would try to hold on, pray that mind would work over matter and that this would allow enough time to elapse for things to calm down so that I could get to the medical experts at Mengo. At the same time I realized we were lucky. Had I not been so close to giving birth, we could have been on the roads near to Kampala when fighting broke out and could easily have become the unwitting victims of violence. Or we could have been stranded on a field trip, or Cliff could have been away and unable to get back.

THE ELUSIVE BABOON

Meanwhile, chaos reigned, and rumor followed upon rumor. We saw open-backed lorries carrying young men who stood in them holding rifles, which they pointed unsteadily outwards giving the appearance they were poorly trained, anxious, and afraid. Fearing they would shoot at the slightest provocation, we stayed off the roads as much as possible and only ventured forth when necessary to go into town to buy provisions. Thus we existed in a perpetual state of anxiety but, considering all that was happening, felt relatively safe on the estate.

That feeling was about to be shattered.

CHAPTER 26

An Unexpected Threat

AFTER FIGHTING BROKE OUT IN KAMPALA and the curfew was imposed, people retired to their homes by seven in the evening. From then until six in the morning the only sounds came from the incessant strumming, humming, and clicking of insect wings and, occasionally, the harsh, strangulated cries of a hyrax echoing across the valley. Fragrant scents of frangipani and the night-flowering jasmine filled the tropical night air and wafted through the open windows. No vehicles drove along the road winding through the estate.

But on Friday, May 27, at around ten in the evening, the silence was broken by the noise of a car turning into our driveway. Cliff had his nose in a scientific book. I had been enjoying Evelyn Waugh's *Scoop* and laughing at the eccentric and beautiful socialite, Mrs. Stitch, at whose feet men groveled. We both looked up. I put down my book. Someone was out there and apparently coming to our house. Yet no one was going out at that time of night. I started to chew my right thumbnail and breathe more rapidly.

From where we sat, we could look out through the French windows and, as the engine noise grew louder, could see the side-lights of a vehicle twinkling through the leaves of the trees lining the driveway. Then the car drew to a halt, and the engine cut off. Silence followed, during which I felt my heart beating faster with the mounting tension. A car door opened, followed by a click as it shut.

Then heavy footsteps crunched on the gravel at the side of the driveway and headed towards the verandah. The beam of a torch flickered as it lit the path of whoever was out there, and before long the figure of a tall, fairly thick-set white man appeared at the French windows. He had a receding hairline, wore horn-rimmed glasses, and was clad in a khaki shirt and long trousers. He stopped, peered at us through the windows, tried to turn the handle, and, finding it locked, pointed to the door with the index finger of his right hand.

We had no idea who he was and held back. But after he rapped sharply on the window and said he had some important news, Cliff opened the doors. The stranger strode inside and stopped in front of me as I struggled to my feet. Wasting no words in pleasantries, he got to the point. He said he lived on the estate and was sorry to come at this late hour, but he had come to warn us there was danger ahead. A tingle of fear spread down my spine, and my jaw tightened as I continued to chew at my nail. Cliff motioned for the man to sit down, but he refused, saying he hadn't much time and still had a number of houses to visit.

He went on to say that trouble was about to erupt on our estate. I gasped and sank in a heap onto the sofa while the men continued to stand. Cliff frowned, pushed back his hair, and asked how the man knew about this and what exactly he meant. Our visitor replied that Norman had told him some disturbing news.

Norman was an ex-U.S. army USAID official who lived on the estate and owned a short-wave radio. He was our main source of information and kept us all up-to-date on events in Kampala.

Now that he had our full attention, our visitor continued. Norman had been visited that day by the Gombolola chief, who was head of the village in the area. According to the chief, the local people were expecting incidents to occur on the estate that weekend, because the Ugandan army intended to get together as many firearms as possible by looting the homes of Europeans. Then the stranger dropped his bombshell: "It is likely," he said, "that you will be shot in your beds. You need to get away."

I felt as if my guts were about to drop out. Five minutes before, we had been preparing for a good night's rest and reading with enjoyment. Now we were faced with this. "But what have we done?" I gasped.

"You don't have to have done anything. You just happen to be in the wrong place at the wrong time," the man replied dispassionately.

I began to suspect he was enjoying the drama of the situation, that he liked being in a position of importance with information that others didn't have and was gratified when his words evoked such horrified reactions. He went on to say he was warning half of the estate about the danger, while someone else was warning the other half. I asked if he was sure of what he was saying, because I couldn't believe that someone would shoot us to get firearms.

He confidently retorted he was quite sure and added we should pack some essentials and leave for Kampala first thing in the morning, on the assumption we would *not* be attacked in the night. He said he had lived in the area for years and knew "these people," by which he meant the Africans. He added that we should trust his judgment and claimed the chief would never say there would be an attack if it were not true. We had no phone, no radio, and no other means of communication with the outside world, and it was dangerous to leave the house during curfew. With no means of checking the accuracy of his claim, we didn't know what to believe.

Having scared us out of our wits, our visitor said he must go. He left, pulling the door shut behind him. We immediately locked it. His footsteps faded as he retraced his path. Then his car door opened and clicked shut; the side lights came on, the engine revved up, and the car backed down the driveway to the road.

I looked to Cliff for some reassurance and asked if he believed our visitor's story. Cliff said he didn't know what to think, but that the man had seemed very certain about what he said. I was puzzled by why he'd come at night, especially at such a late hour, if Norman had heard the news during the day. Cliff speculated that Norman might actually have heard the news in the early evening before curfew, and that the man had been late because he visited others on the estate first. Both of us thought, however, that no one would risk coming out after curfew and at such an hour unless the situation was extremely urgent. It was already Friday night and the beginning of the weekend when the attack was expected. We needed to act quickly. But I still questioned why the man had come when he did. He must have known we couldn't leave in the middle of the night, so why not wait until early in the morning to warn us? Cliff said it was because he'd claimed we might be shot in our beds in the night. If we survived, we must *leave* first thing in the morning.

I struggled to take this in. "Did he mean we could be shot *tonight*?"

"Well, it sounded like it."

I couldn't believe it. Something didn't add up. Why would the army come in the night? Surely the man was exaggerating. But what if he wasn't?

I was about to give birth at any minute, and when I laid my hand on my distended abdomen the unborn child kicked vigorously beneath my fingers. Then I thought of the innocent two-year-old sleeping peacefully in the small bedroom at the back. Pictures of our own country flashed through my mind when I thought about parents, friends, and colleagues all ignorant of what was happening to us. Tears began to well in my eyes at the questions that raced through my mind. How could this be the end of us? How could the children be denied a future? What would our relatives at home be told?

This had to be a nightmare from which I would wake to see the familiar white walls of the bedroom. I would hear people talking in the distance as they walked along the road beyond the hedge. There would be the sound of insects and the musky sweet smells of the mimosa and jasmine. An occasional car would go by. Friends would drop in. I turned to look at Cliff. His shoulders slumped, he pushed back his hair and let out a sigh. There was an expression of great concern and worry on his face. At that moment I realized it was no dream.

CHAPTER 27

Panic Strikes

WHAT HAPPENED NEXT SEEMED FUNNY in retrospect but was terrifying at the time. As the sound of the car faded into the night, the inevitability of the situation took hold, and I became strangely focused, as I had that evening in Genoa. It was a matter of life and death, and I wasn't about to give in. I said we must do something to protect ourselves. Cliff agreed, but the problem was what; there was nowhere to hide from anyone.

I refused to accept this. We *had* to find a place to hide. I insisted we should search around. We went through every room, looking for a place where we would be out of direct range of someone who put a gun through the window. It was no good. The search confirmed what I already knew but had refused to admit; every room was at ground level and had large windows. There was only one place where we were not a sitting target, but this consisted of a space only about four feet by six in the passage-way between the bedrooms and the bathroom.

Cliff said we would have to squat there. Faced with death or discomfort, the choice was clear. We roused Caroline from her bed, and she started to cry as Cliff carried her through half-awake. Somehow the three of us managed to squeeze in and curl up. It was extremely uncomfortable on the stone floor in that tiny, cell-like space. I kept shifting my nine-month- pregnant body around until, after about half an hour of sheer agony, I shifted yet again and said I was sorry but I could stay there no longer.

Cliff helped me up and took Caroline, saying he wished he could do more but it was impossible to do anything to get us out of there while the curfew was on. He thought we should pack a bag of essentials, ready to make a quick getaway in the morning. I stretched my aching limbs and sat down on the bed next to Caroline, who fortunately went back to sleep. Having packed some gear for an emergency, Cliff said we might as well get back into bed and take our chances—that way we could get some rest. If we heard anything, we should be able to get out of the way quickly and into the hall. At least we knew that we had a place to hide if necessary. He also said he would get out his gun and put it under the bed, where he could easily reach it. If necessary, he would shoot.

At any other time, that would have been laughable. As far as I knew, he'd never shot at anything except a target on a range, and I had little confidence in his chances of successfully defending us. The gun was a single-shot 0.22 rifle for collecting baboons by shooting them when necessary, although he personally never used it for that purpose.

Deeply disturbed, we climbed into bed and put our two-year-old between us to protect her, but rest and sleep were vain hopes. Sleeping fitfully, we lay and listened for noises and watched the window, expecting a gun to appear at any minute. Every so often, one of us got up and paced the floor. Cliff kept checking on the gun. Unaware of the danger, Caroline slept peacefully. The minutes and hours gradually ticked by, and I began to imagine noises that weren't there but nothing happened.

Morning came, heralding a beautiful day. The sun shone from a clear blue sky, and there were light breezes. All was quiet. Nothing appeared to be amiss. But we were shaken and tired by the events of the previous evening and our sleepless night. Neither of us wanted much breakfast, and as we sat at the table while Caroline ate, I bowed my head and whispered from the corner of my mouth, "When do you think we should leave? Where will we go?"

"I'm not sure. Probably it's best to head for Kampala, find out what's going on, and whether we can stay with someone."

"I suppose so. There doesn't seem to be any other alternative."

Suddenly a helicopter flew overhead. Its rotors whirled loudly as it circled around, hovered, swooped down, and then up again. I felt my heart pound and cried out, "What's happening *now?*"

Cliff patted the back of my hand. "It's just a helicopter—nothing's going to happen."

"How do you know?" I was so nervous I'd begun to have wild thoughts about bombs and invasions. Childhood memories of World War II came flooding back with alarming and unexpected intensity. However, Cliff was right: Nothing happened. Eventually the helicopter flew off; there was no explanation of its intrusive presence.

Trapped in my large body, I was unable to move rapidly even if I needed to. With our essential items packed, we had to make a move. We had been told to head to Kampala but had nowhere to stay and decided that, before we rushed off, we would review the situation with the Walkers. Maybe we could figure out something together.

Cliff carried Caroline as we crept along the sides of the road, sticking closely to the bushes in the hedgerow to be seen less easily. Keeping our heads down we scurried along like a couple of beetles, glancing around for signs of danger, though we usually strolled down the road at a leisurely pace while Caroline skipped along in front. Anyone watching must have wondered what on earth we were doing, but no one was around.

We soon reached the Walkers' house, hurried down the slope of their front garden and round the back, and, entering the shelter of their verandah, headed towards the French windows and looked inside. To our amazement, they were having breakfast as if nothing had happened, eating their usual toast, drinking their usual coffee, chatting and getting ready to light up cigarettes. Their four-year-old son was sitting on the floor, working his way through a bag of tooth-rotting sweets. His mother was reading a local paper. Apparently they hadn't heard the news.

"It looks as though we're going to have to tell them," I whispered.

Rikki looked up and spotted us. She waved her cigarette and motioned us to come in. "What the hell are you *doing* here? What's the matter with you? You look *terrible!*"

"So would you if you hadn't slept all night. Haven't you heard the news? Haven't you packed?"

Alan blew out smoke and joined the conversation. "What on earth are you talking about? Of course we haven't packed. Why are you here so early?" Neither of them seemed too pleased to have been disturbed.

"We've packed, ready to leave," I said.

They suddenly looked interested. She asked, "Have you gone into labor or something?"

"No. Thank goodness. Didn't you get the message about people on the estate being attacked over the weekend? Some man came around late last night and told us we would probably be shot in our beds and that we must leave for Kampala today. He said we had to pack and be ready to leave early. We stayed up on alert, as he said something might happen during the night. We wanted to make sure we were not shot at in our beds."

The words had tumbled out and I began to feel rather embarrassed by this admission when I saw their astonished faces. Cliff backed me up. "Jen's right. That's what he said."

Alan laughed. "You don't want to believe that nonsense. Norman told us last night there might be trouble but it was unlikely. He said we shouldn't move, as it's as safe here as anywhere."

They both treated the whole thing as ridiculous and continued to eat.

"But we're all ready to go. We were told everyone was leaving the estate," I stammered.

Alan took a sip of fresh brewed coffee, stubbed out his cigarette, and said, "Don't be so daft. Of course everyone's not leaving. Sit down and have a cup of coffee."

I was almost hysterical by then, not knowing whether to be relieved or terrified. "How can you sit there in that casual way as if nothing has happened?"

"Well, nothing *has* happened, has it? So sit down if you mean to stay! You're getting worked up over nothing."

Caroline clung to my skirt. Obviously disturbed by the exchange, she began to suck on her lower lip. I couldn't believe what was happening. There was no way they could have been so unconcerned if they had heard the news delivered to us. Although their words calmed me, I also began to feel hopping mad, because, if they were correct, we had spent the night in terror as a result of some idiot creating unnecessary panic. We had been unable to eat, and here they were tucking in and reading as though this was like any other day. On top of it, they refused to take us seriously.

I turned to Cliff. "What should we do? Should we leave for Kampala anyway?"

"We'd probably better go, Jen, especially as you're about to pop at any minute."

"Pop" was the right word. I looked like an over-inflated balloon.

"Please yourselves," said our friends. "We're staying here." They went back to their breakfast.

We headed back up the road, this time walking down the middle, and got ready to leave. Having loaded our bag into the Rover, we headed into Kampala.

At the last minute, we'd nearly stayed, but it seemed better that I should be near the hospital in an emergency. Cliff left me with friends in Kampala while he went back to Katalemwa when he heard he had to hand in his gun. Later that morning, he and several other people on the estate who owned firearms were escorted into Kampala by the local police. The guns were taken from them and locked away in the central police station. Overall, it was a relief to know we didn't have to fear the gun being stolen and used against us.

In the current state of uncertainty and under the threat of war, people were helping one another. Friends in Kampala generously said we could stay with them, and we moved in on May 28. That evening, we heard the Royal Air Force was standing by in Aden ready to airlift out all British nationals in Uganda. Suddenly I realized how nerve-wracking the situation must be for our parents back in England who had to read about events in Uganda and not know whether we were safe.

"It's a relief to know the RAF will be at hand," I said, "but I don't want to fly out. It would be terrible to give birth on an airplane. Can you imagine the headline, 'Woman Gives Birth In Mid-Flight'? That's not my idea of a moment of fame."

Cliff stared at me and said that he had been thinking of how to get me and Caroline out. He thought this might be the answer. I said I wasn't going. It was more risky than staying. Anyway, where would I go in England? Where would I have the baby? What about Caroline if I went into labor? I said we would sit tight.

The weekend went by. We had been told to expect trouble on the Monday, but nothing of major significance happened, although we heard that the bodies of the two surveyors killed the previous week had been brought Mulago for embalming. I stayed indoors while Cliff traveled around Kampala. He saw cars being searched; his own was searched several times as well. Once he was stopped at the Gayaza roundabout, where the car was searched for grenades by an aggressive man with a rifle, but he was let go. Cliff saw cars riddled with bullets and the tires shot up. Caroline's teacher, who was a reliable source of information, told him she had seen lorries loaded with dead bodies being driven away from Mengo hill the previous week. And yet official reports claimed that only a few were dead.

"What am I going to do? I'm supposed to give birth in Mengo Hospital," I said to Cliff.

"Look on the bright side. It's lucky you weren't there last week!"

"Yes, I know, but we don't know when things will settle down."

My anxiety continued to increase, but by the Wednesday, June 1, things had quieted down, and I still showed no signs of going into labor. We returned to the estate to find nothing had happened but discovered that everyone at our end of the estate had left for Kampala. All of them had been visited by the man who visited us. Everyone at the other end of the estate, including our friends, had stayed. They had been visited by Norman. When I saw first-hand how readily people believed stories when there is uncertainty, especially when the person, like our visitor, claimed expertise and people had no means of testing the truth of what they were being told, I was reminded of the panic created by Orson Welles in the "War of the Worlds" radio broadcast in 1939.

And so we settled back to our usual routine. Our main concern now was for our *shamba* boy, Fred, who was nowhere to be found. Lawrence claimed Fred had taken Lawrence's bicycle during the weekend when we fled to Kampala. "He went to purchase a shilling's worth of beans, madam, and has not been seen since." Lawrence, the eternal optimist, was afraid he might be dead.

Jimmy, our trapper, and the voice of reason, said, "Fred has gone home. He was married on June 4."

We hoped this was true but still worried, because the army had been going around the villages, shooting people. This time we feared "Eeyore" Lawrence, with his focus on doom and gloom, might be right, and that Fred might have been an unwitting victim. Meanwhile, my long wait to give birth continued. It was time for something to happen.

CHAPTER 28

Giving Birth

IN THE MIDST OF THE FIGHTING, the baby came due. I'd had a very difficult pregnancy with a lot of sickness during the first six months, and I had gained too much weight. Uganda coffee is some of the finest in the world, but I couldn't tolerate the smell. Cornflakes or porridge became the basis of my diet in the mornings, though I'd rarely eaten them before. Starting out model-thin, I now looked like the fat lady in the circus. I was tired of the weight, lack of balance, and the same old maternity clothes that, having come unprepared, I'd borrowed.

My due date was towards the end of May and beginning of June, by which time the heavy fighting had abated, but the curfew was still on, and it was not safe to go out after seven. This meant that, each evening, I tried to decide whether I would get contractions during the night and whether we needed to make a dash to the hospital before the curfew struck. With no way to predict when I would go into labor, and deprived of any control over this impossible situation, I was stressed to the breaking point. I imagined all sorts of pains, but nothing happened and the days passed.

THE ELUSIVE BABOON

My casual attitude about the ease of having a second child had long since disappeared. Now I began to suspect there were problems with Caroline's birth in London that no one had told me about and wondered if there was something I needed to know that was relevant to this one. I thought about how we had driven for hundreds of miles over stony ground riddled with potholes, and how I had braced my body against the shock by putting my feet on the dashboard and slipping back on the seat. I worried that I would I produce a tense, stress-shocked, overactive offspring as a result, and I worried about my swollen hands and ankles, with the threat of edema that the doctor had described. Exhausted, I could do little more than lie on the sofa in our cool living room. From there I looked through half-closed eyes at the moving shadows cast upon the verandah when the sun filtered through the leaves that were blowing in the slight breeze. Often I daydreamed about pedaling on my bike along country lanes in England, with the wind rushing through my hair, as I inhaled the scent of wildflowers in the hawthorn hedgerows. Or the quiet shade of the bluebell woods carpeted with their deep-blue flowers in the spring, and where purple violets, delicate white anemones, yellow aconites, and pale-yellow lesser celandines grew at the sides of the woodland path. Or I would think of tobogganing down snowy slopes on winter days, or freezing in the crowd at a soccer match before rushing back to a warm fire and the tempting smell of hot scones and butter. Then, for a while, I forgot my current state.

When I was two weeks overdue with no sign that the baby was ready to be born, the doctor decided to induce labor. So I prepared a small bag of clothes and toiletries, and on Monday morning, June 13, we all piled into the Land Rover and set off along the Gayaza road to Mengo Hospital, wondering what to expect. We had heard the disturbing news that bullets had been pumped into some of the hospital walls during the height of the fighting, and at one point a bullet had whizzed through the operating room. Maybe there was a potential for more fighting. This, together with the uncertainty of the procedure, fueled my anxiety. If the procedure didn't work, what would happen next? Yet on balance, I was relieved. I didn't have to decide each evening that I might go into labor and needed to head to the hospital before curfew. Nor did I have to live in fear of traveling along dangerous roads after curfew and being shot if we decided to head to the hospital. And I felt enormous relief that a professional would be delivering the baby rather than some untrained person trying to do it in the middle of the night and having to deal with any complications that might arise. I trusted the hospital experts would take care of me and looked forward to the difficult pregnancy coming to an end. Everything was out of my hands. All I could do was resign myself to my fate, cling to the knowledge that the ordeal would be over before too long, and hope everything would be fine.

Mengo exuded a reassuring calm. Kind nurses met us and put me to bed before Cliff and Caroline left and the fun started. First they tried "natural" methods, which meant soapy water was stuffed up my rear end with enemas. This did nothing to shift the baby, and I was left feeling sore and disgruntled. The next day, further attempts were made, with no success. That day I was visited by a sweet little nurse named Isibia, who told me with a nervous giggle that she was a good Christian girl and liked nursing. She confided that she hoped the good Lord would help me to have a nice little baby. I thanked her but thought I needed more than the good Lord to help with the birth. I was feeling fed up and sore and wondered what would happen next.

Next occurred on the morning of June 15, when I was taken into the disinfectant-smelling operating room for a surgical induction. The waters were broken, and I was given another enema for good luck. If that didn't do the trick, I was in trouble and dreaded they would be forced to operate. Then the nurses left me alone in a darkened room on a flat delivery table and told me to ring a bell if I needed help. During that long afternoon, as my body finally began to go into labor, I noticed a bullet hole in one of the walls and realized I had been very lucky indeed to have avoided all the shooting. Cliff came with

Caroline later in the day. She stayed with the nurses until they had to leave to get back to Katalemwa before the curfew. And then, just as curfew struck at around seven, contractions began in earnest.

A nurse who was in the room for the birth suddenly gasped and shouted, "Get the doctor. We have an emergency!"

Another nurse came running in, and I heard footsteps rushing out of the room. I didn't remember any pain and didn't feel worried. I was curiously detached, as though I was an observer looking through an opaque screen at an event that wasn't happening to me. Then I became aware of confusion, with the doctor rushing in, rolling up his sleeves, delivering the baby, and thrusting it at the nurse. He told her to wrap him and give him to me and began pumping on my abdomen, saying, "He's alright; it's the mother that's the problem."

I looked at the bundle in my arms and finally knew it was a boy. He looked in good shape and was quite alert. I looked at him with pleasure, but he regarded me with a scowl on his little red face as if annoyed about being ejected from the comfortable surroundings in which he had lodged for the past nine and a half months. Meanwhile, the doctor continued to pump away until eventually I heard him say:

"That's it. We've done it."

He let out a huge sigh, straightened up, and wiped his arm across his forehead. His face was strained, his shoulders drooped, and he looked exhausted. Turning to the nurse he said, "You can look after her now," before he patted my arm and left.

I felt as if I had climbed to the edge of a precipice, fallen off, and been caught in a safety net. I was physically and mentally drained, but I had survived, and a sense of great weariness descended over me. The nurses scurried around, cleaning up. They took the healthy baby from me, swaddled him, and took him to the nursery. They washed me down, put me in clean clothes, and combed my tangled hair. Once I was refreshed, they transferred me to a gurney and wheeled me through to a room where they put me into bed, propped me up with pillows, and fed me light soup and a welcome cup of tea. After that, I slept peacefully for the first time in weeks.

The following day I learned that the doctor had been in the middle of his evening meal and left everything to take care of me. Fortunately he lived nearby. His name was Hugh Oliver, a wonderful Englishman who was caring, knowledgeable, and kind. There were no words to fully express my gratitude to him, and I never really had the chance to thank him properly for everything he did for me.

For the next ten days, I was kept in Mengo to get over the birth. As I lay in bed, I thought about my good fortune to be in the best maternity hospital in Uganda. Founded in 1897 by Sir Albert Cooke, a British-born medical missionary, Mengo had a long and prestigious history and was the oldest hospital in Uganda. Cooke was forward-thinking, devoted to his profession, and known for training Africans to become skilled medical workers. Together with his wife, he opened a school for midwives at Mengo, where Lady Katherine Cooke was matron and the General Superintendent of Midwives from 1897 to 1911. They were well ahead of their times. Ironically, the hospital with its bullet-ridden walls was located on Namirembe Hill; the name Namirembe means "full of peace" in Luganda.

I spent some of the most pampered days of my life in Mengo. Breakfast of fresh bacon and eggs, paw-paw, and mangoes was brought to me in bed, and I had the room to myself. What a pleasant contrast it was to my experience in a London hospital when Caroline was born and I was in a room with about nine other women. Every morning we were hauled out of bed at some ungodly hour and made to sit on a hard bench for breakfast. Most of us perched gingerly on the edge because our non-soluble stitches, mine from a painful episiotomy, felt like barbed wire underneath. Breakfast was cold bacon, eggs, and beans held together in light-brown congealed fat. A small, rather brawny nurse with short dark hair and a voice like a sergeant major would take her position at the head of the ward, peer over the line of beds, bark commands at us, or complain because the bedclothes were not turned back the requisite number of inches, which I thought was ridiculous. She seemed devoid of compassion, and I dreaded it when she was on duty.

My hospital room in London had no windows. The light was provided by a dismal electric bulb, and as there was no clock, it was hard to gauge the time of day and I become disoriented. I thought prisoners must experience this when confined to a cell with no daylight. At Mengo, windows and doors led from my room to a balcony. The doors were opened during the day to let in the smell of fresh air, the scent of blossoms in the lush garden below, and the faint buzz of insects. In the early morning I could see little valleys filled with pockets of mist from which velvety-green hilltops emerged. In the garden below the balcony I saw a hibiscus with bright red flowers that looked luminous when the sky was overcast, and a pomegranate with delicate leaves and beautiful orange-red flowers that would drop and be replaced by the yellowish-red fruit. Shafts of sunlight poured into the room in the late afternoon, and the skies turned gold, orange, red, and yellow at sunset. In this idyllic setting I was very comfortable.

The English staff nurse was the one minor drawback. She would come in the morning, perfectly turned out in a navy blue skirt and blouse, and a starched white apron. A small white nurse's cap perched on top of her dark hair. She was probably about my height and age. Compared with the relaxed style of the African nurses, she was businesslike, strict, and at times overly bossy, but occasionally she stopped and chatted about England and said how much she missed it. Then I felt sorry for her, but unfortunately her other side mostly predominated.

One morning she took my temperature and pulse, looked up from her chart, stared intently at me from the foot of the bed, and remarked, "If you have another baby, you'll have to go into hospital, you know."

I was eating my breakfast and stopped mid-chew. I had no idea what she meant but I could feel alarm signals spreading through me. ". . .No, I didn't know. No one said anything."

"You had a PPH."

"What's that?"

"A post-partum hemorrhage, of course." Apparently I was supposed to know this.

She continued, "You lost lots of blood. You're lucky to be alive."

Suddenly my breakfast didn't taste so good. I put down my knife and fork, leaned back against my pillows, and stared at her. Finally I found my tongue. "Well, I knew there was a problem. I didn't know it was that bad."

"Oh, yes, you lost a lot of blood."

"Was it really very bad?"

"Of course it was. The placenta wouldn't come out. Lucky for you the doctor was there to handle it. Otherwise you could have bled to death."

Now I was really put off my breakfast and wondered whether she was exaggerating. If not, I hadn't realized until then how dire the situation had been and that the doctor had saved my life. It was a humbling experience. Meanwhile, the nurse hadn't finished with me. I was pondering what she had said when she brought me back to earth by adding, "You've got terrible stomach muscles now!"

Maybe she was trying to be helpful, but her comment seemed gratuitous to me. I almost laughed. What did she expect, an immediate return to a smooth svelte-like figure?

"Yes," she continued with relish, "you'll need a lot of work to get into shape. You may never get your stomach muscles back. You're in a bit of a mess."

I was dumbfounded. Maybe I was too sensitive, but I resented being told I was a mess even if she meant well. I felt my face become expressionless. I was fuming but didn't want to say something that I would later regret, so I guarded my tongue. "Well, I suppose I'll have to work on them, then," I managed to say through tight lips as I clenched my fists and took a deep breath.

THE ELUSIVE BABOON

After she departed, I surreptitiously lifted the covers to examine the offending region. It seemed to me that things were not so bad, considering I had just given birth. Admittedly, I looked a bit wrinkled and fleshy, but what could you expect after being stretched over nine months, putting on thirty pounds, and giving birth to a baby weighing nine pounds and eight ounces?

The head African nurse, Sister Sali, had known the Kabaka and, like the other Baganda, held a deep affection for him. She told me the Kabaka had escaped from Uganda to London. His supporters were sure he would return, and his people would fight without regard for their own safety to establish him in his former position. Various members of his household had fled, and one or two of his most loyal servers had managed to accompany him out of the country, but others had been killed.

Like many of the local women, Sister Sali had an ample, motherly figure and was compassionate, caring, and kind. Generally she was cheerful, but one day she was very serious. She explained the nurses were extremely concerned about women in the outlying villages who needed professional care, and even Cesarean sections, but were unable to get to the hospital because of the lack of transportation or danger from the army. No one knew how to find the women and transport them safely into Mengo. They feared some of them would die. We had heard rumors of the army killing people in villages and knew Lawrence had a boy staying with him whose aunt had suddenly disappeared from her village, leaving behind five children. Sister Sali confirmed these rumors. But the reality of my own situation struck home when I heard about the potential fate of the pregnant women. Had the baby come at night, and had we relied on someone untrained or risked the roads, the outcome for me could have been completely different. I had been extremely fortunate.

During the two weeks when he was overdue, my baby had developed considerably. At night the nurses would apologetically get me up to deal with him, because he refused to be quiet. I would go into the nursery where there were several African babies neatly swaddled like small sausages in blankets from which their little black heads stuck out. They lay quietly in their cribs. But at the end of the row was a crib containing a large white baby who was kicking off the sheets. This baby had healthy lungs and demanded attention. He yelled when I tried to put him down after feeding, and I discovered he was objecting because he wanted to look around to watch what was going on.

Cliff brought Caroline to the hospital to see him, and she was thrilled. For a start we were unable to reach agreement on a name, but finally decided on Erik. Before then, Caroline informed the nurses she had a baby brother called "It". He was a rather windy baby, and I was amused to hear her telling the assembled nurses that this was because "It" had cold feet. She was allowed to visit, and the nurses took a fancy to her. They sang nursery rhymes, and I once heard them all singing, "Baa Baa Black Sheep." When Caroline informed them she intended to hold the baby, they said she was too small, but she was very offended, saying, "I not, I'm big."

One nurse kept teasing her saying, "That's my baby," to which our two-year-old retorted:

"No, it's not. It's mine, so buzz off." She was starting to get a bit too cheeky, but no one seemed to take offense, and the nurses even encouraged her by roaring with laughter.

The fighting had died down, but the political situation remained unsettled. However, while I was in the hospital, we had an excellent piece of news. Fred was alive. Cliff took Lawrence to look for him and brought him back with his new wife, Robin. On the way, they were stopped twice, and the Land Rover was searched, but they were eventually let go. Gradually we pieced together Fred's story. He had left the estate to get married and had set off for his village on a bicycle borrowed from Lawrence. He took his wages and the extra money we had given him for his wedding. On the way, some ruffians stopped him, hauled him off the bicycle, beat him, robbed him of his possessions, and left him badly hurt in the bush. He lay there for some time before he managed to make his way on foot to his village, where he was helped by friends and eventually recovered from the attack.

Robin and Fred were wonderful when I came home. Robin was a charming, pretty, intelligent girl with a round open face and high spirits who helped me with the children. I guessed she was about sixteen. She spoke very little English, and I was unable to speak her language of Luganda, but we seemed to communicate reasonably well. She was extremely anxious to talk to me, never gave up trying, and would chatter away in Luganda using sign language. She was always laughing and cheerful and a great pleasure to have around. Showing lots of initiative, she coped well with the children. However, I think the time when we really communicated occurred quite by chance when I happened to be singing.

Usually I was too self-conscious to sing alone, but one day I had chosen a song I had not sung in many years and never sang again. Thinking no one was around, I let rip, "Come follow, follow, follow, follow, follow, follow me. Whither shall I follow, follow, follow, whither shall I follow, follow thee? To the greenwood, to the greenwood, to-o the-e greenwood, greenwood, tree." Suddenly I realized I was not singing alone. A voice in the kitchen was singing along with me, and then Robin appeared with a big smile on her face. It turned out that she had learned the song in a missionary school. We both laughed before she went to look for Caroline who was causing mischief.

CHAPTER 29

Wrapping Up

HOW DO SOME WOMEN MANAGE to produce ten or more children and keep going? Why do some bounce back so quickly after childbirth? Giving birth left me exhausted in a way I could never have imagined. I felt like a limp rag doll. In spite of the wonderful rest in Mengo, I could hardly drag myself from room to room when I first returned to Katalemwa. Robin had arrived at just the right time, and I constantly gave thanks for her calm, cheerful presence and help with the children as I began to recover my strength.

Like most babies, Erik woke at night to be fed and sometimes kept me awake for hours with his fussing, so that I walked around like a zombie during the day. Early on we discovered he loved company. If I tried to put him down to sleep in another room while we ate dinner, he kicked up such a rumpus that we brought him into the dining room and propped him up in his basket at the table. Immediately he was quiet, and his inquisitive little eyes moved around as if taking everything in. Though demanding, he was good natured, and by August was smiling and whooping with pleasure, especially when he rode in

THE ELUSIVE BABOON

the Land Rover.

After the enforced break in his research due to the war and Erik's birth, Cliff was determined to make the most of the time we had left. The political situation seemed to have settled down sufficiently for him to go on field trips. On one occasion he flew to Nairobi for a few days and while there visited Amboseli National Park, where a well-run and well-funded team of Americans was trapping baboons to send back to the States for medical research. They allowed Cliff to take blood from their baboons. Amboseli turned out to be a good place to work because there was a shortage of food, which meant the animals could easily be lured into traps. It also became clear that a large, well-coordinated team was essential for successful trapping.

Cliff continued to collect vervet monkey blood from different sources. Some of it came by train from Nairobi with peculiar messages warning of its arrival. Once, while he was in Budongo, a cryptic telegram arrived that read *Blood and bodies sent passenger train today—Millervakvet Kabete*. Another time someone took a message and wrote down that blood was coming from Vera Muscovitz in Nairobi. I wondered who this woman was with the mysterious Russian-sounding name. It all seemed excitingly like James Bond, and I was disappointed when it turned out to be no more than a problem in recording the message. Each time blood arrived, I had to make sure it was picked up and properly stored at Makerere.

People would appear at the house with dead or live monkeys, and at one point we had one monkey in a cage on the verandah while another occupied a cage in the spare room. I ended up having to feed the monkeys and numerous rodents, in addition to dealing with the children, when Cliff was away, and I was very glad to have the help of others. On one of these occasions the monkey inside got sick, and although Rikki tried to revive it with milk and whiskey, the wretched thing was very thin, had diarrhea, and stank. Terrified that it was carrying disease, I wanted it nowhere near my children. Alan saw it, decided the kindest thing was to put it out of its misery, and to my great relief dealt with it.

Then Jimmy arrived with two men. One was carrying a live monkey, the other a dead one. They put the live one in a cage outside, where it immediately ripped off the front and leaped away, followed by a band of Africans who had gathered in our garden to see what was going on. It escaped up a banana tree. After that, the little monkey on the verandah escaped when Cliff was away, and I had to get several Africans to help catch it.

On top of this, Caroline started to cause trouble. Initially she was delighted with her baby brother, but the novelty soon wore off, and she became jealous. She decided she was sick of Erik and demanded that we take him back to the nurses. When she realized she was stuck with him, she got her revenge by waiting until I was busy feeding him to cause mischief. One afternoon she went to the pantry, found a tin of cocoa, went to the spoon drawer, prised the lid off the jar of cocoa, proceeded to empty half the contents onto the verandah, covered herself in cocoa, and sprinkled the rest around the house. I told her to go to her room to play with her toys. All was quiet until she appeared wearing a cloth nappy around her shoulders and slopping along in my shoes. Later I discovered she had found a box of hairgrips, taken out a sheet of sticky labels used to label baboon and monkey specimens, and affixed them all over the grips. Another day she emptied Jeyes Sanilav (a toilet cleaner) over the bathroom floor. She put the remains of a toilet roll down the toilet and pulled the chain; the package of corn flour I used for cooking disappeared because she said the ants had eaten the contents; and the bathroom plug went missing.

One day she told me that if I didn't take Erik back to the nurses, she would stick him in the dustbin. She took every opportunity to pummel the poor little chap on his stomach when he was looking around quite happily, and in the evenings, when I wanted her to go to bed, she would head off to Fred and Robin and hide under their bed, where I couldn't reach her. To top it off, she locked herself in the bathroom with the keys, and we had a terrible job to get her out.

THE ELUSIVE BABOON

Often I felt rushed and overwhelmed, but I also knew that my life overall was easy and pleasant. People were always around to help. Neighbors visited or came for meals, or we went to them. When necessary, women on the estate assisted one another with their children. Robin babysat if we went to visit neighbors in the evening, and Caroline loved to spend time with her. I was never bored, and as our departure for England grew closer, I began to realize that, although we had been through some difficult times, I would miss this life when I got back and had to manage on my own.

Shortly before we left, a lovely American family from the Midwest, the Goodmans, arrived on the estate. John was a pleasant, funny, kind man who reminded me of Bob Newhart. He was a herpetologist. Not until he arrived and offered money for captured snakes was I aware that some deadly varieties lurked in our garden. Fred captured one from a tree and took it on the end of a forked stick to John's house while people scattered and squealed as he headed down the road. John said it was some sort of bird snake that was highly venomous. In his own garden he found a snake which he thought was a Boomslang. Someone else claimed it was a harmless Jackson's tree snake. But John consulted with a friend said to be America's foremost herpetologist. Together they decided it was definitely a Boomslang, and a very deadly snake. To drive home his point John told me one had killed K.P. Schmidt in 1957, when Schmidt was one of America's foremost herpetologists.

One morning when John visited, he said he was having difficulty getting the Africans to understand him. He had decided he could make himself more intelligible by pronouncing words like English people, but this caused even more confusion. I wasn't surprised. Even I couldn't understand his efforts. He had gone to purchase a can of beans and pronounced can "carn," thinking the English would broaden the "a". I had to laugh. Not only did the English *not* broaden the "a," they called a can a tin, and so did the Africans. Little wonder there was a mix-up. He also said he had gone to purchase ten fifty-cent stamps and spoken each word slowly to the man behind the counter: "Ten fifty—cent stamps, please."

Several thirty-cent stamps were produced.

"No, feeerfty, feeerfty."

"Oh," said the man, "feefty."

"Okay. Now I want tern, tern," John repeated.

The man frowned as if working something out then suddenly smiled, "Ah, tin, sir!"

"Yup, I guess so."

Among the visitors who came and went, the most notable was the physical anthropologist Dr. Nigel Barnicot, who had taught Cliff at University College and had been the instigator of the baboon project. He was the person who said Cliff's tiny writing gave him a pain in the neck. When I first met him, I thought he looked like Socrates with his bushy beard and snub nose. Based on little information, I had formed the impression of a rather austere individual who wore a plain blue tie, a navy blue suit, and puffed on a pipe, so when Cliff announced, "Old Barnicot's coming to visit and stay with us," I had some misgivings.

My views changed as soon as Nigel arrived casually dressed in khaki slacks and an open-necked shirt. He was relaxed and friendly and said that he was meeting up with a colleague working at Mulago Medical School. The two of them were heading off to collect blood, earwax, hair, and various other unmentionable secretions from the Hadza people living on the Serengeti Plain. Nigel said they were going to sleep out on the Serengeti. Cliff told him he'd heard stories of hyenas coming and eating people in the night, but Nigel just laughed it off.

He proceeded to prepare for his trip at our house, spreading his equipment all over the living room in carefully assorted piles. A large canister of liquid nitrogen to preserve blood samples sat in the middle of the floor. This fascinating liquid starts to evaporate in swirls of white mist like the graveyard scene in a horror movie once you take the lid off a canister. We had been told it could be dangerous and should be treated with respect. We told Nigel, but he took no notice. On the morning of his departure, his colleague picked him up in a Jeep, and Nigel climbed into the passenger side with the canister situated behind it. I had visions of it exploding and ejecting him into the air like a human cannon ball, but he was unconcerned and waved cheerfully as they set off, saying, "Not to worry! Cheerio! See you in about four weeks."

Three and a half weeks later, and somewhat earlier than expected, his whiskery face appeared at the door. He hadn't blown himself up or been eaten by hyenas but had had a very successful trip measuring the Hadza and collecting vast numbers of blood samples and other bodily specimens. The Hadza were delighted because they were well paid and highly entertained by the whole process.

Cliff was still away in Budongo, and Nigel started to follow me round the house, talking of his expedition. I was just about to bathe Erik in the kitchen sink, because this was much easier than the bathroom, when Nigel came in and plunked a large bucket of raw meat in the sink before he started to sort through it. He informed me it contained a mixture of guinea fowl and dik-dik for us to eat. Nigel was oblivious to the fact that I was about to bathe Erik in the sink. I was speechless. Luckily, Robin and Fred came to help, and I sponged Erik down in the bathroom. Having dumped the meat, Nigel proceeded to take out numerous vials of blood which he spread out on the kitchen counters and dining room table in order to label them. There was blood all over. Cliff returned the following day, and the two of them had a long conversation about the stuff Nigel had collected while I cleared up and cooked the meat.

Nigel and his wife, Nina, had no children, but he became friendly with Caroline, and together they discussed the animals in Beatrix Potter's books, with which she was very familiar. He confided that his favorite animal was the big fat rat, Mr. Samuel Whiskers, which she agreed was an excellent choice. Nigel presented us with an enormous bone from the leg of a rhinoceros, and a disintegrating monkey skin, when he departed. However, he gave Caroline a delightful small toy rabbit in a blue-and-white spotted dress, covered in a white lace-edged apron, which he had thoughtfully brought from England. It became a great favorite. In spite of his eccentricities, Nigel turned out to be an engaging and entertaining visitor who seemed happy to chat and pleased to stay with us. We were sad when he left and afterwards sorely missed his company.

Finally, it was time to return to England. Rikki and Alan drove us to Entebbe Airport, where our flight was delayed because Obote was meeting Kenneth Kaunda, the President of Zambia, who was flying in. I felt deep sadness on leaving friends, the congenial atmosphere and the wonderful country. But we had also missed family in England and the changing seasons with the red, orange, and yellow hues of autumn leaves, the freshness of spring, and a cold winter to kill off bugs. We had had some exciting and interesting times in Uganda but had also suffered a lot from illness, and it was time for a break.

So, exactly a year after our departure from London, we boarded a VC 10 to return. I had never flown before and found my first flight an exhilarating and magical experience. I fastened my seat belt, leaned back, and thought about what lay ahead, until my drifting thoughts were broken by the roar of the engines. The plane began to gather speed on the runway, and then I felt that strange sensation of leaving the ground when you seem to hover in space before climbing up and up. From my window seat I looked back. The tarmac glittered in the sunlight. On the hillside, the painted letters signifying *ENTEBBE Airport* were receding rapidly in the distance, and Lake Victoria's surface sparkled against the brown-and-green countryside we had come to love. I knew we would be back.

African houses near Katalemwa

Waiting for the ferry

Land Rover on ferry with baboon strapped on top

West Nile hunters

Caroline with children on Mount Elgon

Women transporting goods on the road

Camping at Pakwach

Camp at Budongo

PART TWO: RETURN TO UGANDA

"Men laugh at apes, they men contemn.
For what are we, but apes to them?"
—From John Gay's *Fables*.
"Fable XL: The Two Monkeys"
(John Gay, 1685–1734)

CHAPTER 30

Amin Seizes Power

AFTER WE LEFT UGANDA IN 1966, Milton Obote continued as President. The Kabaka had fled into exile in London with some of his loyal supporters after being overthrown by the Obote regime. Sadly, this once wealthy and revered man ended up poverty-stricken and died in 1969 under suspicious circumstances in London's East End. He was only forty-five years old. Officially his death was ruled a suicide from alcohol poisoning, but many believed he was murdered by agents of Obote who force-fed him with vodka. John Simpson, a BBC reporter who interviewed the king earlier in the evening, said he had seemed cheerful and there had been no signs of him drinking. His death remained a mystery. In 1971 his body was returned to Uganda and given a state funeral.

In 1971 we had the opportunity to return to Uganda as well. Since our last visit Cliff had taken a teaching job at New York University, and we'd left England in late 1967 as part of the brain-drain to live in Manhattan's Greenwich Village in university housing. It was a time of much turmoil in the United States, including protests, with cities looted and burned. During this time of civil unrest in the States, the *New York Times* reported that tension was developing in Uganda between President Milton Obote and Idi Amin who was by then a major general. Amin had been Obote's Army Chief of Staff, responsible for the ousting of the Kabaka, and was involved with Obote in the 1966 gold scandal. Now they were rivals in a bid for power. The problems between them escalated. In January 1971 Amin got wind of information that Obote intended to arrest him on charges of misappropriating millions of dollars in military funds. He retaliated. While Obote was out of the country attending a Commonwealth conference in Singapore, Amin staged a military coup on January 25 and seized power. He was said to be supported by Britain and Israel.

Amin then declared himself President of Uganda, Commander in Chief of the armed forces, Army Chief of Staff and Chief of Air Staff. Obote was reputed to be in Tanzania preparing for a comeback. The two men flung abusive comments at one another. Amin accused Obote of corruption. Obote said Amin was the greatest brute an "African mother has ever brought to life." Meanwhile, the Ugandans were said to be tired of the Obote regime and pleased with the change to Amin. But the *New York Times* reported there was a lot of tension and general unrest after Amin took control, with fighting in Kampala. We followed events carefully because we had to decide whether it would be safe to travel to Uganda that summer.

We knew very little about Amin except by reputation, but stories portrayed him as someone who could be dangerous and unpredictable. Born around 1923, he had grown up in northwest Uganda and had barely any education, but at age eighteen, during World War II, he had enlisted in the British Army and in 1946 joined a British colonial regiment, the King's African Rifles. He did well in the army, was promoted to colonel and army commander during the overthrow of the Kabaka in 1966, and by 1968 he had been promoted to major general. An imposing and formidable military figure, Amin stood six-foot-four, wore size 13 shoes, and weighed over three hundred pounds. Physically strong, he saw himself as omnipotent, and as a god on earth who was said to have written to the Queen of England and various heads of state in garbled and inappropriate language. For instance, he was said to have addressed Her Majesty the Queen as "Liz." He offered refuge to President Richard Nixon of the United States after the Cambodia invasion and, boasting of his accomplishments, sent missives around the globe.

Sexually promiscuous and proud of it, Idi Amin Dada or "Big Daddy," had fathered numerous legitimate and illegitimate children. He was larger than life, ate and drank to excess, and no one dared question or cross him for fear of torture or losing their lives. He saw himself as lord of all the beasts of the earth and fishes of the sea. Some said that he was insane but no one knew what to do about it. A former heavyweight boxing champion, he asserted he could beat Muhammad Ali and claimed he had challenged Ali to a fight but that Ali had been too scared to face him. However, he was said to be pro-British, which was in our favor.

But with the situation so uncertain, we monitored the political reports carefully. As the time for our proposed departure drew closer, things appeared to have improved. We decided to go. Nevertheless, we developed a contingency plan. We would fly to Kenya, rent a car in Nairobi, and drive to our destination on the edge of the Budongo Forest in Uganda. There we would stay in an old rest house at Busingiro used by visiting scientists. If trouble broke out in Uganda, we believed we could beat a hasty retreat to Kenya. The plan had the added advantage of allowing us to see more of Kenya and combine work with pleasure.

This time we were going for two months to collect samples of foods eaten by baboons, to have them analyzed for content. The project was far less ambitious than the previous one of trapping animals to take blood and arose from a debate among physical anthropologists about tooth proportions in primates and its link to human evolution. In a nutshell, the question was whether alterations found over time in the teeth of fossil hominid remains were more likely to have been due to a change of diet than to the use of tools.

Scientists knew the front teeth in fossil hominids had been reduced relative to the back ones. Some argued that this was due to tool use claiming that if tools were being used in food preparation, there would no longer be a need for strong, long front teeth to tear food apart. Cliff disagreed. In a seminal paper, "The Seed Eaters," written in 1970 at the relatively young age of thirty-one, he put forward the view that the front teeth were very slightly reduced while the back teeth grew in order to grind down hard grains that were a staple food. Furthermore, he argued that seed-eating was linked to development of the opposable thumb to hold the seeds, the need for primates to squat to eat the seeds, and, ultimately, that this was linked to the evolution of upright walking in man. His paper was hailed in the academic community as a creative breakthrough in physical anthropology because it departed from the traditional view. But at that time little was known about the kinds of foods actually being eaten to begin to substantiate his hypothesis. By collecting and analyzing the foods consumed by baboons, he hoped to shed some light on the issue.

Before we could go, all of our vaccinations had to be up to date. Caroline was then seven and a half and Erik almost five. He was an extremely active child who had an unfortunate habit of jumping off the kitchen counters, closet shelves, and some large concrete blocks on the plaza surrounding our apartment buildings in Manhattan. His nursery school teacher complained he was teaching other children to do the same, and that it had to stop. He was always running away, and I constantly chased after him. I suspected much of his behavior was due to the bouncing he had experienced before he was born in 1966, when we traveled around Uganda in the Land Rover during my pregnancy.

He also seemed to have an innate ability to figure things out, and at the age of two had taught himself to stand on blocks to unlock the door to the apartment, and how to climb on the kitchen counters and turn off switches in the fuse box over them, to plunge us into darkness. He would hide whenever

THE ELUSIVE BABOON

he didn't want to do something, and this happened when we went to get shots for yellow fever. The ones Caroline, Cliff, and I had had six years earlier were still valid. Erik was the only one who needed a shot. As soon as he saw the doctor bring out a syringe with a needle on the end and realized he was going to be the lucky recipient of its sharp point, he did an amazing disappearing act under a large table, shrieked horribly, and refused to come out. No amount of cajoling or threats worked. The doctor was irritated having to wait while we struggled to extract Erik kicking and screaming. Then we had to hold him down for the doctor to get the needle in before we scurried away, feeling acutely embarrassed.

Finally we were ready, and towards the end of May, soon after the pink cherry blossoms formed a mass of color in Central Park, we set off from John F. Kennedy Airport. As usual we took the overnight flight to Heathrow and, as always, I was struck by the early morning mistiness of London—the soft green countryside and the damper, kinder, somewhat musty air formed a sharp contrast to the harsh atmosphere and brilliance of the sunshine against concrete in New York. We visited relatives near London before Cliff departed for Nairobi and I followed some days later with the children. Another exciting journey was about to begin.

CHAPTER 31

Into the Highlands

BEFORE AIRLINES BEGAN OFFERING discounted fares directly to customers, the only way to find affordable fares was to take a charter flight. So we had visited a small, dingy office strewn with papers and coffee cups on London's Charing Cross Road, where a young woman with a cigarette dangling from the side of her mouth sat behind a gray metal desk and squinted at us from behind heavily framed glasses. The place looked disreputable but offered cheap flights. We told her we wanted the cheapest ones she could find for me and the children. Taking sips of coffee and blowing out cigarette smoke, she leafed through various documents until she found a cheap charter flight. We took the risk, bought tickets, and left. On June 21, I got to Gatwick Airport with the children and, because booking the tickets had seemed so dicey, was greatly relieved when a plane showed up. But things were disorganized. We had not been allowed to book seats in advance, and I had to fight to get on line to obtain seats with the children. This was followed by a long wait during which the captain announced in a cheerful Irish brogue the disconcerting news that we would be delayed "due to instrument trouble." But we eventually took off late that evening, with dinner served at midnight, and the trip itself was smooth. At around ten in the morning local time, the familiar letters ENTEBBE, site of Uganda's main airport, appeared etched in the hillside.

We were to land, unload mail, and refuel before flying on to Nairobi. As we touched down, the flight crew announced we could briefly disembark at Entebbe but warned us we must not take any photographs of the Ugandan army, air force, or police. No explanations were given, though I was alarmed to hear someone say there were rumors of war. Several of us climbed out to stretch our legs, emerged into sultry air, and stood on the glistening tarmac. A propeller-driven craft used to transport police, and three small planes with the legend *Uganda Air Force*, sat on the runway. For the first time I felt uneasy about being there and held tightly to the children but told myself that the planes were signs of Amin's recent takeover and not an indication that war was imminent.

As soon as the plane re-fueled we took off and landed at Nairobi International Airport forty-five minutes later. A long, straggling line of people stood with cardboard boxes, suitcases, brown paper packages, and small children who, like mine, were moaning and protesting in the airport's stuffy, hot interior. One weary-looking person checked all the health cards while others sat to the side, chatting and doing nothing to help. It took a frustrating ninety minutes to go through customs and immigrations.

Cliff met us in a borrowed car. He had enjoyed himself during our absence because he had flown north in a small plane piloted by Richard Leakey to visit the archeological site at Koobi Fora, near Lake Rudolf (now Lake Turkana) in the African Rift Valley. Richard was following in his famous father's footsteps and hunting for fossils hominids. On that visit they found fossil bones but no human remains.

Driving north toward our hotel, we skirted the edge of the National Park.

This was an area of about 30,000 acres covered in short grass and small thorny acacia bushes. It housed some wonderful wildlife including giraffes, lions, zebra, warthogs, gazelles and baboons. Then we headed through central Nairobi, where tropical trees and bougainvillea of intense reds, oranges, purples, and pinks grew in profusion at the sides of Kenyatta Avenue and Uhuru Highway. Eventually we arrived at the Ainsworth Hotel. This old colonial hotel was situated on the northern outskirts of town close to the National Museum, where Richard Leakey was the director. Reasonably priced and very clean, it provided good service and courteous staff. Central Nairobi boasted luxury hotels like the New Stanley, but they were expensive and, anyway, we preferred ours. However, Nairobi was growing very quickly, and it was hard to believe the bustling modern city had been little more than a shanty town when it was originally established in brackish African swamp as a depot in 1899 during the construction of the railway.

We unpacked, went outside, and sat down for coffee. Although the weather was cool, it felt fairly comfortable. Around us the brilliant purples, pinks, and reds of bougainvillea grew in abundance in the hedges; huge spiked euphorbias grew in the garden, and the scent of exotic flowers drifted through the air. Overhead the sky was pale blue, and in those calm surroundings we were able to relax and rest from the continual noise and bustle to which we were accustomed. It was so different from London's damp grayness and the fast pace and glitter of New York.

Next to the hotel, an open scrub area bordered the road. A group of six men dressed in khaki shorts, shirts, and rubber sandals sat on it as they methodically chipped flakes off uneven stones with hammers and chisels. They worked at a leisurely pace, chatting and laughing as they made blocks for a building being erected behind them. Compared to mechanical devices there was little noise, but having just come from the United States, where such work would have been done by electrically powered machines, we stared in astonishment.

THE ELUSIVE BABOON

Before I arrived, Cliff had purchased a car from a second-hand dealer, and over the next two days we made final preparations while the car was overhauled by the museum's mechanic courtesy of Richard. At around 8:30 in the morning of June 24, we went to pick it up. By then we were eager to be off and pleased to see the garage mechanic grinning as if he had good news to impart. Rubbing his hands on a grease-stained rag, he advanced towards us in navy overalls spattered with oil. After we exchanged greetings, he beckoned us to an open lot where the car, a rather battered-looking, beige-colored Ford Cortina, was parked. He stopped, turned to us with a big smile, and to our dismay said cheerfully, "It's a terrible old wreck! Held together with wire and rubber bands when I got it. It will probably get you there and back as long as you treat it carefully. Treat it as you would an old lady, and don't knock it around as you would a young one. Handle her with care and respect, and you should be all right." Cliff seemed to find his comments amusing, but I had been exposed to Women's Liberation in New York and didn't take kindly to the sexist remark.

Disregarding my scowl, the mechanic gave a chuckle, returned to another old vehicle on which he was working, and left us with our "wreck." In spite of his dire warning, we were delighted to have transportation. Keeping our fingers crossed that he had done a good job, we piled into the vehicle, drove back to the hotel, loaded up the car, bought food, confirmed our flight back to London, and took the main road going in a north-westerly direction out of town.

On the outskirts of Nairobi, young teenage boys sold sheep-skin rugs and hats by the roadside. Many of them wore hats to demonstrate how they looked, and the effect of these white, furry coifs above an outfit of khaki shirt, shorts, skinny legs and bare feet was rather like that of a furry duster on a stick. A boy would leap into the road, holding a rug in front of him like a Spanish bullfighter with a cape, while a group of his friends stood at the roadside, encouraging him with shouts and laughter. Further off the road, the "back room" boys pegged skins on rope lines. They were scraping and de-fatting the raw skins then tanning them and combing the fleece into a fluffy mass. We stopped to look and said we would purchase something on the way back but felt sorry when the faces of the enthusiastic salesmen fell.

Our Cortina had suffered at the hands of rough drivers and bumpy roads. The price had been right, but it refused to travel comfortably above 45 miles per hour. In spite of this, we soon cleared the city and were in good spirits. Our plan was to reach Kampala in two days, breaking the journey somewhere in the Kenya Highlands around Eldoret.

Chugging along steadily, we crossed the southern tip of the Aberdare Mountains and began to drop down into the Rift Valley, with the foothills of the Aberdare Range and the railway to our right. After about thirty-seven miles, the distinctive volcanic cone of the dormant Mount Longonot jutted out from the surrounding flat landscape and rose into the clouds on our left. The sharp ridge of the Mau escarpment loomed in the far distance. Our route through the valley took us through rich agricultural land along a straight, flat road, so that in a relatively short period we reached Lake Naivasha, a center for wildlife and the birds found in abundance on its shores. On we went to Gilgil and to the soda Lake Nakuru, where we stopped to look at the beautiful brilliant pink and red flamingoes for which it was known, and which spread out in every direction around the lake. We had seen these lakes from a distance when we took the train journey in 1965, but to view them close up and see the birds massing on their shores was a totally different experience, for the colors were quite breathtaking. We didn't linger, because we planned to spend ten leisurely days on the way back viewing all the sites and so kept going towards the highlands.

The car gallantly plodded along at a slow pace as the road began to climb until we passed through rich agricultural land that had been extensively farmed with maize, wheat, beans, and other crops since big farms were established during colonial times. Then we went over the equator and on to Eldoret at an altitude of about seven thousand feet above sea level. The journey was uneventful, and we saw few people.

Eldoret had been established around 1910 in the midst of farms created by white, mostly Afrikaner, settlers who came up from South Africa after the Boer War. It had a post office and was known as "Farm 64" because it was sixty-four miles from the new Uganda Railway. It was officially named Eldoret in 1912 and grew after the Ugandan Railway extension reached it in 1924. Mostly it was an administrative center and, when we arrived, quite small. We didn't stop but decided to head to Soy, about fourteen miles further on, where someone had recommended we should break our journey at a hotel. Driving north, we came to a sign pointing off the main road to Soy and Kitale, turned onto it, and were excited to see four huge giraffes lope gracefully across the road ahead of us with their necks rocking back and forth to the rhythm of their movement. And then we came to a sign pointing left to *Soy*. Cliff swung the steering wheel and we soon found ourselves on a muddy track leading to some African huts among the trees. This was quite unexpected, and we suspected we were on the wrong road but, curious to see what lay ahead, we decided to keep going and to our surprise reached another small track at the end of which we saw a cluster of brick buildings surrounded by a high hedge. We headed towards them, came to a gate, and drew to a halt where we were thrilled to see a sign that read *Soy Country Club*. But our delight was short-lived, for under this ran the legend, *No children under eleven admitted*. We had two of those.

CHAPTER 32

Charlie Appears

WHEN SOY WAS RECOMMENDED, we had no idea it was a country club, and no one had told us children would not be allowed. After the initial shock, we debated whether to go back to Eldoret but realized we might not get there much before nightfall, and there was no guarantee of anywhere to stay. We decided to find out if we could stay at Soy.

As we got out of the car and made our way through a gate in the hedge that led to the club, I felt like Alice going through the looking glass into a totally new world. One minute we had been surrounded by small African huts; the next we were surrounded by low brick buildings, with bedrooms and other quarters, situated on the perimeter of a square courtyard. Unlike the busy hotel we had anticipated, the place was strangely deserted. We scouted around, hoping to find someone, and after few minutes were glad when an African in a white shirt and gray slacks appeared from one of the buildings. He approached and told us he was the manager of the property. We explained our predicament; he looked down at the two children clinging to our hands, told us to wait while he consulted with the lady who owned it, and disappeared. A few minutes later he returned to say we could stay and directed us along a track at the back of the club where we parked the car close to our room. There seemed to be no other visitors.

We had just settled in when the owner appeared. She was of medium height, probably in her fifties, and wore a mid-calf floral dress I associated with the 1940s. She said that she would send a pot of tea. Her accent sounded upper-class English, but it was not the upper-class we were used to hearing and more like that from a 1930s movie. She said little else and, although courteous, was hardly warm and welcoming.

After about ten minutes, an African in a waiter's uniform appeared carrying a tray with a teapot, cups and saucers, some egg sandwiches, and little cakes. This "proper" English tea appeared to have been produced from nowhere. The man put the tea on a small table set up on the verandah by our room, and when we went outside to sit and enjoy it, we found ourselves overlooking a beautiful velvet-green golf course that stretched away from the back of the compound. Mist was rising over the trees, and it was quite cool, so we needed our woolen sweaters. Around us the air was fresh and clean, silence reigned, and we sat back to relax after a hard drive. It was still fairly light.

Suddenly, a small head with a sharp pointed beak, bright beady eyes, and an arc of bristles around the back of its head, like an overgrown golden-colored crew-cut or a chimney-sweep's brush, popped out from around the corner. On seeing us, the eyes lit up, the rest of the creature emerged, and a splendid crested-crane, Uganda's national bird, stood before us. He had black, white, and grayish-blue plumage on his back; the top of his head was black, with the golden crown at the back. To the side of each eye, he had a white half-circle with a red mark above and one below that extended under his chin. Standing over three feet tall on long, slender black legs, he strutted rapidly towards us in a purposeful manner while we wondered what was going to happen. We didn't have to wait long. On reaching the table the bird extended his long neck, stuck his head forward, grabbed one of the cakes and sent a cup flying. We all jumped back while the creature made off with the cake leaving our little party in a state of confusion. We rescued the cup and tried to regroup, but no sooner had we collected ourselves than he reappeared and, refusing to respond to any threats or shooing off, grabbed another cake. From where we sat, his long sharp beak reached eye level, and we got the impression he was not averse to using it. While we were trying to decide what to do, he was back to snatch an egg sandwich. He moved off eating hurriedly and dropping crumbs as he went. A retinue of small birds swooped down and picked them up while we grabbed the remaining food and ate it.

Feeling satisfied we had beaten him at his game, we thought the bird would come back, see no more food, and go away. He returned, but instead of being put off by the lack of food started to dip his beak in the sugar. We gave up, abandoned him to his task, took ourselves to the wall of the verandah, sat on it, and watched. After consuming a great deal of sugar, the crane moved off. Meanwhile the children went into fits of laughter at the entertainment provided when their parents were frightened by a bird.

Soon after, the lady of the house re-appeared, and I remarked we had been visited by a crested crane. She was not surprised and informed me his name was Charlie. When I commented that Charlie seemed to be a very hungry bird, she replied without bothering to hear more, "Oh, yes, he does like a bit of sugar now and again." Then she turned on her heel and walked off, leaving me open-mouthed and thinking, What about our cakes and sandwiches?

Soy was a fascinating place. African homes were close by, and it seemed to be some distance from any European settlement; yet everything appeared to be ready for a huge party. There was a bar, lounge, dining area, golf course, and lots of bedrooms, plus a full staff. We presumed it was a meeting place for Europeans in the area and a golfing center, but it was impossible to find out much about it from anyone there. Everything was charming, but I felt as though we had slipped back about thirty or forty years to the time of the colonial settlers and were surrounded by ghosts of its past.

While the children explored outside with Cliff, I sat in the baronial living room, with its wooden beams and large armchairs covered in floral-patterned loose covers, and looked into the log fire, which had been lit to provide warmth from the cold evening air. Gazing into the flames, I was overcome with the feeling that I would look up and see a lot of white people congregating for a social occasion. They would be the settlers, the farming people of the Kenya Highlands. The men would typically have reddish-brown, weather-beaten faces, good Harris Tweed jackets, and gray flannels or shorts, with brown-suede brogues and knee-length socks. They would have short-cropped hair, probably moustaches, and might have Scottish or English public school accents. I imagined the women would be tough physically; they would have to be to live in that area, and I imagined them in their floral print dresses, perhaps rather bored, looking for scraps of gossip and discussing problems with help on their farms. The scene came alive in my mind; at one point I broke my reverie and turned, half expecting to see someone, but the room was empty. I turned back to the crackling logs.

A fire had also been lit in our spacious room to ward off the chill, and we dressed respectably for dinner, afraid that if we appeared in our crumpled jeans and T-shirts, we would cause offense. We were the only people in the dining room, but the place looked as if it had been set for a reception. The white table linen was spotless, the crystal sparkled, the fire glowed in the grate, and there was a small bowl of beautiful flowers on each table. The African waiters, dressed in white jackets and black trousers, were formal and courteous. Dinner was superb, consisting of delicious freshly cooked beef and vegetables, followed by dessert. It was the equivalent of eating in one of the best restaurants in New York, and yet we were in the mountains and far from any cosmopolitan city. No one spoke to us. That night we fell asleep to the sounds of frogs, in an ornamental pond, croaking in the chill air.

When we left early the following morning, we saw only Charlie the crested crane, and an African. We didn't see the owner, and I began to think she had been a figment of my imagination but Charlie certainly wasn't. Once more he caused problems when we were packing the car, ready to depart. The children were busy collecting some delicate and intricately woven weaver bird nests, which had fallen out of a tree, when Charlie appeared. He advanced in his confident and purposeful way towards the children, who then did the worst possible thing—they hid the nests behind their backs. Charlie seemed to think food was in the offing and began to advance, staring with his sharp beady eyes. Fortunately, an African helper who saw what was happening flapped a cloth at the offending bird and shouted, "Cha-lie, Cha-lie." Charlie stopped in his tracks to ward off the cloth and we took advantage of the pause to push the children to safety in the back of the car. At this, the bird decided to stick his beak in the air and stalk off, as if in a huff. He was the only crested crane in the vicinity and seemed to consider himself the lord of the area around the garden, where he was treated with great deference by the smaller birds, but we never found out where he came from or why he was there.

The air was misty, dew was thick on the grass, and it was cold at that elevation. We set off and retraced our steps. Within a hundred yards of leaving the golf course, we were transported back to the African homes in among the trees, and then to the rough track. We drove along it to the turning, joined the main tarmac road, headed westward, and started to descend to the heat of the plains at Broderick Falls.

At about mid-day we reached the border with Uganda, filled out forms, and went through Customs to the other side. Soon we recognized the volcanic plug standing out as a landmark near Tororo, the vegetation became thicker and lusher, and we gradually came into more densely populated areas. Near Jinja, the fields of sugar-cane had huge mechanical water-sprays twisting round and continually roving over them. Then we came to banana country and the

THE ELUSIVE BABOON

familiar stalls piled with tomatoes, pineapples, paw-paw, oranges, and passion fruit. Modern buildings began to appear ahead, and we knew we were approaching Kampala.

Our main concern on reaching the city was to find somewhere to sleep, so we headed to the Speke Hotel, where we had stayed on our first visit, for it held pleasant memories. But the prices had risen so much they were prohibitive, and we ended up at the Hotel Equatoria. This suited us well though I was a bit taken aback one morning when I saw dead ants on my breakfast bacon. I called the waiter over and told him there were ants on my food. I expected an apology but he just smiled and said in a reassuring tone, "Don't worry madam, they're dead," then left me to it. A lot had changed since 1966. Outside the Grand Hotel, where we used to see boys with baskets trying to sell handmade wooden animals such as rhinos and wart-hogs, stood a covered market labeled *Uganda Crafts*. There were many new and expensive buildings, and far more tourists than we had seen before. This surprised us, especially if it was true that the country was as unsettled and dangerous as some reports indicated.

I was also reminded of the warning at Entebbe Airport not to take pictures when I was sitting in the car by the Kampala market. People swarmed around asking to have their photos taken, and I was doing this when a man suddenly came up to me, shaking his fist, glaring, and shouting, *"Hapana. Ingiri picha."* At his vehement outburst I immediately put the camera away, though I wasn't sure why there was a problem. Later we found out that people who took photographs could be arrested as spies, and that it was important not to behave in any way that might arouse suspicion. The seriousness of this was reinforced when we heard unsettling rumors circulating about the army killing people for no good reason. The situation had been frightening when Obote took control in 1966, but for most of our stay Kampala was calm and we could move around freely. Now it appeared to be far more dangerous, and we heard robberies were not uncommon. Knowing this, our goal was to get out of the capital as soon as possible. However, we had to stay for two days to collect the necessary gear for camping, have the car serviced, buy food supplies, and wait while there was a ceremony for an old chief who had died, before we left for Masindi and the Budongo Forest.

Since 1966 the roads had been much improved for tourists and, as we left Kampala, the tarmac was far more extensive, so that we traveled quickly to Kibanga, where a new bridge was being constructed over the River Kafu. Here we left the smooth road and went onto the familiar red *murrum* but still progressed well until we were almost at Masindi, where a heavy thunderstorm had flooded the road and turned it into a mud bath. Cars had tried to get through, but the road had been churned up and had become extremely rutted. Several cars had skidded into the ditch, and while we waited, several more impatient people skidded in. There was no way our little car could get through, and we stayed until a tractor came along to clear a narrow space for everyone to move forward.

Beyond Masindi on the Butiaba road we passed through an area which we remembered had been intensely cultivated with sugar, coffee, and tea plantations on our last visit, but very few remained, and the land was becoming overgrown with scrub.

By late afternoon we were about thirty miles from Masindi skirting the Budongo Forest. Cliff said we were almost there. No signs appeared, but when we came to a break in the trees on the left, he slowed down, and we saw a track wide enough for a four-wheeled vehicle winding its way up a hill. We turned onto it and followed it slowly upwards for about a hundred yards before it took a sudden sharp turn to the right. About ten yards further on, we found ourselves in an open area with a building situated halfway up the hill on our left. We stopped. Cliff told us to get out and look and then said, "Well here we are. That's Busingiro House. Pretty good isn't it?"

My mouth dropped open.

CHAPTER 33

Busingiro House

BUSINGIRO. THE NAME WAS SOFT AND LILTING. It flowed easily off the tongue and sounded like warm wind rustling gently through the leaves of shady trees. Cliff had stayed at Busingiro Rest House during our previous visit to Uganda. He said it was basic but spacious with a beautiful garden and spectacular views. In its day it had been elegant, with amenities such as hot water and room service. But it had been abandoned and fallen into disrepair. Two years before we were to visit all of the window-shutters and doors had been ripped off. Since then, Makerere had had the place done up for use by research scientists.

No one explained "done up," but that hadn't stopped me from filling in the gaps. I knew the doors and window shutters had been replaced, and we were told two rooms had been "reserved" for us. To me that had the ring of a hotel, and I was not disabused of this when we arrived in Kampala and heard the rest house contained furniture including a double-bed, tables and chairs, that it had a beautiful garden, and that someone was taking care of the property. Even better, this must mean it would be all ship-shape and ready for us when we arrived. I conjured up grandiose pictures of sitting on the verandah in the evening and relaxing with my favorite gin and tonic. It would be like a vacation I looked forward to.

When we stopped on the drive and Cliff pointed to the house, I was dumbstruck. Never could I have imagined what I saw. My first impression was of a wreck in a terrible state of disrepair and appeared to be little more than an abandoned shell about eighty feet wide. It had been built in 1930s Bungalow style, with walls of red brick that looked solid, but the corrugated iron roof was rusting badly. There was no sign of habitation, although someone had zealously attacked an area sloping down in front of the house, which presumably was the garden. The slaughter appeared recent, because a bougainvillea had been chopped down and left on the ground, and a stray group of bright purple flowers still waved bravely at the end of a branch. There were tufts of coarse grass, but the rest of the area was lumpy, bare, and stony. A few stunted bushes remained.

I was shocked and thought, What the heck is going on here? but, having come all that way, decided I had better take a closer look. Maybe things would not be as bad as they seemed. To hide my disappointment, I managed a noncommittal, "I see. Maybe we'd better go and look inside the house."

I bundled the children into the car and jumped into the passenger seat. Cliff commented on my lack of enthusiasm and was obviously disappointed but got in, shifted into first gear, and moved slowly up the hill. We took the last bend, pulled up at the side of the house, and got out to inspect. The building seemed intact, and we saw the promised doors and window shutters. A large verandah stretched across the front, which we found to be in good repair when we stepped onto it and advanced towards two French windows in the middle. They were the only way to see in, because all the window shutters had been closed behind chicken wire. I dragged the family over and looked through the glass into a very large room darkened by the shuttered windows. It was completely bare.

THE ELUSIVE BABOON

No one appeared, so we decided to go inside using a key we had obtained at Makerere. After jiggling it in various locks, we eventually opened a door to the side of the property, which led into a dim and somewhat gloomy room with chinks of light seeping in from cracks in the shutters. It too was completely bare, but looked more cheerful when we opened the shutters and the sun came in filtered by dust motes. We also found three small rooms leading off one another. One led to the enormous room into which we had peered from the verandah. We'd left nose prints on the glass. This room stretched across most of the house. A poster advertising the Toronto Argonauts, and two small pictures of snow in Canada, left by previous occupants, hung on the wall.

Two doors leading off the main room were locked. Another contained an old white enamel bathtub. Some pipes came out of the floor, but we discovered there was no running water. The tub appeared to have been hauled out of its original position and dumped in its present spot. We went through the accessible rooms, opening shutters, and the place became more welcoming with the glow from the sun, but apart from the bath and the posters, it remained completely bare. I could imagine my father's voice saying, "Well, this is a rum how-do-you-do. There's not even a bloomin' sausage!"

Delighted to be away from the confinement of the car, the children raced around gleefully, but I continued to worry about the lack of furniture. I hoped it was in one of the locked rooms, and that someone would soon appear to help out. Meanwhile, we went outside to look around.

Close to the back of the house, two huge tanks gathered water from the roof. They had taps towards the bottom. This appeared to be our water supply. A small path led from the back door down a slope to a brick building with two rooms. One contained the remains of a broken-down ancient brick oven and was black with smoke. The other showed no signs of use. To the side of the house, an earthen path led to a small wooden building. Inside was a lavatory, or toilet, as Americans would say, with a wooden seat and lid. Beneath it, a deep drop led to an earth pit. Flies buzzed around, and I was horrified when

Cliff warned us to put the lid down to keep bats out. I was appalled to think of perching there and having a bat fly up from the hole to poke me on my bare behind. Built when this part of Uganda was scarcely touched by "Western" technology, Businginro House had no electricity, so we had no refrigerator and no electric light. Never in my wildest dreams could I have imagined we were coming to such a place. Everything was a far cry from our New York apartment with its large refrigerator and dishwasher, and two bathrooms containing showers, baths, basins, and toilets.

In the back we stumbled over stones and hillocks hidden in the vegetation and found ourselves under constant attack by large swarms of sweat bees. Although annoying, they were not harmful. We discovered they lived in little waxy tubes that adhered to, and stuck out from, the walls of the house.

We wandered to the front and saw a young African man, dressed in khaki shorts and shirt, and probably in his late teens, pushing a bicycle up the steep slope of the drive. As he drew closer, he gestured towards the house and to himself, pointing backward and forward. We gathered he was the person responsible for taking care of the property and were delighted to see him. He came to a stop in front of us and we exchanged the usual greetings, "*Jambo, habari, ah, ah, eh*," and so on. He said his name was Marco. I was trying to get over my initial shock and dying to find out about the furniture. "Marco, there is no furniture in the house. Where is it?"

I was greeted by a puzzled frown. "What is it, Madam?"

I immediately jumped to the erroneous conclusion that communication was going to be difficult and decided to accompany my words with hand signals, pointing and contorting the parts of my body most affected by an article in question. As I did this, I noticed my children's faces. They probably looked like mine when my mother embarrassed me on school sports days when she appeared in one of her stylish hats adorned with various feathers, pompoms, or fancy pins, while the other mothers wore wooly hats or headscarves. I'd wanted to disown her. Now the children and Cliff looked as if they wanted to disown *me*. Marco waited patiently and looked surprised by the acrobatics, but after I finished said, "Ah," pointed to the house, and indicated we were to follow.

I turned to my family. "What are you all grinning about? After what I did, he seems to know what we want."

They ignored me, and I knew they were thinking he would have shown us anyway, that there had been no need for the pantomime.

Marco rested his bike against the side of the house and led us inside. He stopped at one of the locked rooms, got out a key, unlocked it, signaled we should wait, went outside, and returned a few minutes later with a metal bar like a burglar's jimmy, which he used to prise the door open. It swung back to reveal a dark interior and, as our eyes adapted to the gloom, we saw a stack of benches and tables. In the middle was a bed. We found some trestle legs and, with Marco's help, dragged out two pieces of mahogany, each larger than a door, and set them up on the legs. One was to be our dining table; the other was for Cliff's work and would hold the foods foraged when he followed baboons. After that, we pulled out some extremely heavy benches, also of fine mahogany. Now we had somewhere to sit, but there were no armchairs. Finally we lifted out the bed, which was much smaller than we expected, and set it up in a large room leading off the main room. After that, Marco departed and said he would return the next day.

The sleeping quarters were a challenge. Although wire reinforcements had been put on the windows to keep out intruders, the glass that should have been behind the wire was shattered or missing. In addition, none of the windows had the type of wire needed to keep out mosquitoes, and they became a major problem. Fortunately, we had brought mosquito nets. We suspended them from the beams that ran across the ceiling, stretched out the nets, attached strings to them, led the strings to nails in the walls, and secured them. The result was a filmy white tent over the bed in the middle of the room. This diaphanous hanging made me think of a sultan surrounded by his harem.

"Bring on the dancing girls!" I muttered.

Cliff just raised his eyebrows, and went off to get the rest of the gear from the car.

Then there were the children. I was afraid to leave them at night in a separate room in the strange house. What if they got up and wandered around in the dark? Who knew what might be lurking around if they ventured outside by mistake? What if we didn't hear them? In the end we decided the safest solution was to keep them in the room with us to one side of the bed. They were to tuck down in sleeping bags on the camp beds we brought with us. They were very excited when we set up mosquito nets over their beds, and the arrangement could have worked had they not kept climbing in and out and destroying it.

We set up a small wooden table as a wash stand under the bedroom window. On it I placed a red plastic bowl we had brought with us and a white enamel water jug we found in the store room. Then I put soap in a dish and arranged wash cloths and towels in blue and white on a string line. The place began to look colorful and somewhat homey.

Nevertheless, my memories of sleeping in that room are not happy ones. The bed was cramped, and it was stuffy under the net. Mosquitoes flew around, emitting high-pitched whines like small fighter planes waiting to attack, and they were merciless if they got inside the net. These small vampires constantly fed at our expense, and one morning I saw a mosquito so distended with blood it could hardly move. As usual, Cliff tolerated the discomfort more easily than the rest of us, but even he had to admit the conditions were not ideal.

Our poor children would roll into the netting during the night. Poor little Erik suffered terribly. He reacted more violently to mosquitoes that the rest of us and became covered with bites and bumps. He was the only one with blood type O, which, we were told, the mosquitos favored. Thus, the nights were uncomfortable, and we religiously took our malaria tablets every day for protection.

But in spite of all the problems, the house grew on me, and I was especially grateful for its cool interior during the heat of the day. I also began to appreciate some of its former glory and was sad to see it had been much abused. A few window fittings remained where curtains had once hung, but holes in the walls signaled that most fixtures had been stripped away. The fireplace in the main room was bereft of most of its surround, which had been of fine mahogany because some pieces still remained, but most were blackened by recent wood smoke. On the first evening I tried to light a fire, but smoke billowed into the room, causing us to cough and splutter so badly we had to abandon the idea. The room with pipes indicated there had once been running water provided by the tanks, but this no longer functioned and we had to haul water in from outside. With no power, we used a hurricane lamp or Calor-gas lamp in the evenings and cooked in the old kitchen over a wood fire. Once settled in, we began to explore further to see if we could find out more about Busingiro.

CHAPTER 34

Stepping into the Past

APART FROM THE GARDEN in front of the house, which Marco had attacked, and an area at the back where the path led down to the kitchen, the property was very overgrown. In places the vegetation was taking over so rapidly you could see it would soon return to the jungle.

To the side of the drive we found a building with a rusty corrugated iron roof. It had once been a workshop, because a cavernous drop had been dug out of the soil, so mechanics could work underneath vehicles. This was extremely dangerous, especially for children. We immediately secured the door and put it out of bounds. At the back of the garden we discovered pits dug for rubbish and two large reservoirs hidden among the trees close to the room with the broken oven. They contained shallow murky water full of frogs and toads, and also posed a danger if a child fell in. There were other abandoned out-buildings, but we found no more danger spots.

Cliff said that, when he visited five years earlier, he had discovered an area with four metal posts like those marking the corners of a tennis court. We went over to the place he remembered, but the posts were gone. However, when we stopped, bent over and inspected the ground carefully, we found to our surprise and satisfaction that we could trace an area equivalent to the size of a court where the ground had been leveled out by raising the soil at one end and cutting into the hill at the other. More clues to its existence came from grass and small saplings that had grown up on the area and were less dense than the vegetation surrounding it. But the land was quickly disappearing into the forest. Unless you knew where to look and what to look for, you would never have known a court once existed. I tried to imagine what it must have been like, with players clad in white outfits racing around, hitting a ball over the net, and shouting cheerfully to one another before going indoors to relax.

Behind "the kitchen," a path wound down between the reservoirs to an abandoned building. Treading to the side of the path, we stumbled over rocks hidden in the grassy slope. We knelt, pulled away thick grass, and found the stones had been arranged to form a rockery. Beyond the kitchen, we pushed through thick undergrowth and suddenly emerged into a shady avenue bounded by mango trees that someone had carefully planted in lines. A building at the end of it contained an abandoned wooden press to prepare sugar from canes.

Brick steps led from the verandah to the garden in front. This appeared to have been terraced at one time, but again, tough grass had taken over. Tall elephant grass grew abundantly around the area slashed by Marco, and the boundaries of the property were hard to figure out. However, we found a track about half way up the drive that branched off into the elephant grass, wound its way up the hill, emerged at the side of the house, and went on through the trees at the back, where it dropped down to a farm belonging to an Asian who grew sugar cane.

The driveway must have been impressive when built. One side was cut into the side of the hill, which went up steeply towards the house. The other side bounded a drop down the escarpment. Large, smooth, oval stones had once marked its edges, but they too were mostly hidden among the invasive grass. The more we searched, the more we discovered, and I began to appreciate some of the excitement that archeologists must feel when they uncover lost civilizations like Mayan ruins in the jungle.

The old water system became one of our most intriguing finds. We knew the room next to the bedroom had pipes that came up through the floor, but there were no taps and no running water. Where the water came from and how it got there was a mystery until Erik came running to me shouting, "Mummy, quick, come and see!"

He dragged me into the small room containing pipes where water had gushed out and spread over the floor. He and Cliff had gone into the trees at the back of the house, where they found a huge open metal tank with a tap. They turned on the tap and returned to the house to find the water pumping out. It must have come through pipes under the garden and into the bathroom. Presumably rain and water from the abandoned reservoirs had supplied the tank.

Cliff said he had figured how hot water had been supplied and led me to a brick construction near the back of the house. We called it the barbecue pit, because the bottom contained some blackened ashes, as though someone had lit a fire there. He pointed to the remains of pipes a few feet from the ground that went through its walls. Above them was a large opening, which he said would hold a metal drum. Water would be directed into the drum from the pipes, and a fire would be lit in the bottom of the pit. This would heat the water in the drum, which would be directed through more pipes into the house. How exactly it worked, and whether it had once been connected to the two tanks of water that now gathered water off the roof wasn't clear, because the whole system was so broken down it was hard to figure out.

The remains of a brick summerhouse with windows on three sides stood about a hundred feet to one side of the house. Yet the only view from there was of trees that grew up against it. We realized these trees, like those on the tennis court, were saplings. At one time the area around the summerhouse must have been cleared to allow visitors to look across the escarpment towards Lake Albert. The rotting remains of a seat around the inside perimeter suggested it had once been a pleasant place to sit, read, reflect, and view the landscape. Now everything was covered in mango droppings left by baboons.

Mango trees grew in abundance and surrounded the perimeter of the garden. One enormous specimen between the main house and the summer house had a long branch to which someone had attached a metal fixture for a swing. We got together some wood and rope to fix up a swing for our children, and this gave them hours of pleasure.

But the spectacular views set Busingiro apart. From the verandah you could look for miles across the surrounding countryside. To the right, the forest stretched into the distance and was at its best in the early morning when small pockets of white mist lodged in among the dark green leaves. Then we heard screeching and chattering as life stirred in the branches. Ahead of us the wooded escarpment dropped down to a flat plain covered in scrub, with small bushes. A thin reddish-brown ribbon appeared every so often marking the *murrum* road that wound its way down to Butiaba on the side of Lake Albert in the Great Rift Valley. Even though it was about twenty miles away, we could see the lake quite distinctly and, on clear days, were able to trace the outlines of the blue mountains of the Congo rising to the horizon beyond. The English explorer Samuel Baker must have been awestruck when in 1874 he became the first European to visit this enormous lake, which he named "Albert" in honor of Queen Victoria's late consort. A hundred miles long and nineteen miles wide, Lake Albert receives the water of the Victoria Nile, flowing from Lake Victoria via the spectacular Murchison Falls. Then, close to the mouth of the Victoria Nile, the Albert Nile flows out of the lake, to become the White Nile, one of the two major Nile branches that unite at Khartoum, and eventually find their way through Egypt to the Mediterranean.

In the evenings, the sun set over the lake and turned it into a liquid gold band so brilliant in intensity that it dominated the surrounding area. Sometimes, spectacular storms broke over it. When these happened after dark, white lightening ripped the sky apart, with jagged arms going in different directions to light up the countryside in an eerie purplish hue; other arms disappeared into the lake when they shot down to the water. Sometimes, sheets of lightning lit up the ghostly shapes of trees and bushes in the garden. And the rain came down in torrents, forming small streams that rushed explosively down the hill, where they eroded soil in their path. At their climax, the violence and intensity of these storms filled me with awe. But on calm evenings when the skies were clear, we could look through our telescope to see the craters of the moon.

As we continued to unearth pieces of the property, I realized Busingiro House must have been a beautiful place in its heyday in the late 1930s, when flying boats owned by Imperial Airways carried passengers and mail between Europe and South Africa, via the Nile corridor. The planes would land on Lake Albert and stop at Butiaba to refuel, and passengers would be driven up to Busingiro to stay over, relax, enjoy the breezes and the company, play tennis, sit in the summer house, and see the spectacular views. I imagined it was a place where stories were exchanged by travelers and where they could sit, eat, and bathe in comfort.

But I wondered how long Busingiro House would continue to exist. It lacked basic amenities, and I predicted it would be abandoned when the scientists decided to stay there no more. I thought this would happen fairly soon, because few people would tolerate such lack of comfort. The local people would then salvage anything of use, and the property would quickly disappear into the forest. But Businigro had aroused our interest, and we wanted to know what else the surrounding area would reveal.

CHAPTER 35

Butiaba and Beyond

BUTIABA, ON THE SHORE OF LAKE ALBERT, had had an interesting past. Most people who had heard of it would probably have associated it with Ernest Hemingway, when he was writing a series of articles for *Look* magazine in the early 1950s. The story was that, in 1954, when he was on safari with his fourth wife, Mary, they had flown over Lakes Albert and George, and the Murchison Falls, when the tail of their light plane clipped a telegraph wire. It crashed onto the banks of the crocodile-infested Nile before settling in the scrub. Mary suffered two fractured ribs, and Hemingway either dislocated or sprained a shoulder. Together with the pilot, they were forced to camp overnight in hazardous elephant country. A launch taking tourists to Murchison Falls found them the next day, picked them up, and took them to Butiaba.

Hemingway was reluctant to fly again but wanted to get out of Butiaba. Mary persuaded him to charter a de Haviland Rapide plane to Entebbe, piloted by Reggie Cartwright. Cartwright's plane gathered speed on the airstrip, hitting ant hills and thorn bushes that caused it to leap like a wild goat according to Hemingway. Briefly it became airborne, then plunged down and burst into flames. This was the second crash in three days. Mary and the pilot escaped through a window. Hemingway, hampered by his size and weak shoulder, was stuck. He finally head-butted himself out of the blazing wreckage but sustained such extensive injuries that headlines in the *Daily News* read *Hemingway Feared Dead in Nile Air Crash*.

That he survived was a miracle. He was said to be badly burned and suffering as well from other injuries that included a ruptured liver and kidney, a fractured skull, impaired hearing, and problems with his vision. A police rescue party took the Hemingways to Masindi, where they received medical attention and spent several days recuperating at the Masindi Hotel. After that, they traveled by road to Entebbe, where the *New York Times* and *Washington Post* reported they arrived on January 25, 1954, with Hemingway swathed in bandages. He carried a bunch of bananas and a bottle of gin in his uninjured arm. In typical swashbuckling style, Hemingway made light of what happened, yet he was still too damaged by October to attend the ceremony when he was awarded the Nobel Prize for Literature. Many believed the injuries he sustained at Butiaba marked the gradual decline in his mental health over the next seven years and ultimately led to his suicide in 1961. It was a tragic end to an incredible story.

But many years before Hemingway arrived, Butiaba had been a flourishing port during the Colonial era. At that time it had a major harbor and became a booming commercial center and transportation hub when a regular steamer service was established in the 1920s to transport merchandise to it from the Belgian Congo across Lake Albert. From there, it went overland through Masindi to Masindi Port, where it was loaded onto barges and ferried across Lake Kyoga to Saroti. By 1929, the railway had reached Saroti, so the merchandise was then put onto railway wagons for transportation to Mombasa for export. Imported goods and merchandise were transported along the same route in reverse. And, of course, Butiaba was an important landing site for the first flying boats that came through the corridor of the East African rift

THE ELUSIVE BABOON

valley in the late 1930s.

One of the boats that shuttled passengers and cargo across Lake Albert was the *S/L Livingstone*. Built in England in 1912 and used by the British East Africa Company, the vessel remained a nonentity until John Huston saw her in 1951 and commissioned her for a film he was shooting with Humphrey Bogart and Katherine Hepburn. The boat became famous when she was renamed the *African Queen*. The German gunboat *Queen Louisa,* which Huston used in the film, was actually a steam tug named *Buganda* that was operating on Lake Victoria. At one time a channel was built to transport goods from Lake Albert to a point downstream of Murchison Falls, and Huston chartered a boat, the *Murchison*, to carry supplies and equipment down there during filming.

Hepburn later recounted the perils they faced in Africa. Most of the crew and cast were sick for much of the filming and suffered from dysentery, malaria, contaminated drinking water, and close brushes with wild animals and snakes. Maybe they had failed to thoroughly boil or sterilize water, which we knew from previous experience was essential in the field. Hepburn had suffered badly from dysentery from drinking the water. I knew how debilitating that was, and thought it was a testament to her fortitude and strong will that she carried on. She never missed a day of work on a punishing schedule in Spartan conditions. Bogart and Huston were the only ones who didn't get sick and claimed they avoided illness by mostly living on imported Scotch whiskey. According to Bogart, "Whenever a fly bit Huston or me, it dropped dead!"

Having heard about its interesting past, we were anxious to visit Butiaba and so headed off one afternoon by car, following a series of long S-bends leading down the steep escarpment of the rift valley to the lake. On the way we came across the foundations of abandoned brick buildings at the sides of the road. They were all that was left of its prosperous past, when the road between Masindi and the port of Butiaba was called the Caledonian Road because so many Scotsmen lived there running large coffee estates and vast tea and sugar plantations. Now it was little more than a landscape of hidden memories, with the invasive scrub taking over. One farm remained, but when we spoke to the owner, he said he too was going to pack up and leave.

At the bottom of the escarpment, the land flattened out into a plain with grass and small bushes spreading out on either side. We drove across and arrived at Butiaba to find it was now nothing more than a poor fishing village. A few palm trees grew there, and a small group of thatched huts had been set up on the sandy beach for the fishermen and their families. Some small fishing boats were pulled up by the lakeside, and, silhouetted against the background of mountains, we saw a man in khaki shorts punting a skiff on the lake while his two companions netted fish. Apart from that, there was little activity, and it was hard to believe Butiaba had once been a bustling center of commerce. So what had happened? It turned out that the once-thriving port had been struck by an unforeseen tragedy of epic proportions in 1962 when unusually heavy rains caused the level of Lake Albert to rise by several meters overnight. All the ships sank, and much of the town was submerged. After that, the goods and equipment needed by plantation owners could no longer be brought in and out of the port, and Butiaba never recovered. It was officially abandoned in 1963. This had a devastating impact on the surrounding area. Many years of prosperity came to an abrupt end and explained the abandoned buildings of the sugar, coffee, and tea plantations that we had seen along the way.

We parked and got out to look for signs of Butiaba's prosperous past. At the end of a long spit of sand the rusted and tilted remains of an old lake steamer, the *SS Robert Coryndon*, named after Uganda's governor from 1918 to 1922, were gradually sinking into the water. An elderly white-bearded gentleman carrying a stick and dressed in a flowing white *kanzu* shuffled past with an old blue seaman's peaked hat perched incongruously on his head. An old Coca-Cola sign hung on the side of a small building. The only other signs of Butiaba's former importance were the remains of a chapel and some brick houses surrounding it. All the roofs had been removed to use elsewhere, and the walls were being whittled down to their foundations, leaving them to stand like ghosts from the past. A metal pole with a wind indicator on top, and the remains of a rough narrow runway overgrown with grass, still existed to form a sad reminder of Hemingway's aborted flight.

People were friendly and curious. Before we left, they crowded around our car, pointed and smiled, but it was difficult to communicate, though the children had no inhibitions about trying even if they didn't understand the language. With his passion for cars, Erik busily explained the various parts of our old vehicle while the crowd nodded as if following every word.

For the next few days we continued to explore. One day we visited the saw mills in Budongo. Mr. Knight, the manager in 1966, was no longer there. New managers were hiring men to cut down mahogany trees to make furniture for sale at large profits. The noise of buzz-saws cutting into trunks, the creaks as they started to give way, and the crash when the trees hit the ground filled the air. Magnificent mahoganies were being rapidly depleted, and much of the wood was being wasted.

We were fortunate to see some mahogany trees still standing. Many were over six hundred years old and rose majestically to the sky. But we also saw some that had been felled and abandoned. One lay across the path with a severed trunk about four and a half feet in diameter facing towards us. Others, for some unknown reason, had their stumps burned black, and they faced the elements like modern sculptures. With heavy hearts we saw how hundreds of years of growth could be brought to a murderous end in minutes. The beauty of the forest was being destroyed along with all the wildlife it supported. Conservationists like our friend Vernon were taking action to try to save this magnificent area, but it was extremely difficult due to resistance from the forestry managers and problems in growing new trees.

One day we pushed through swamps and elephant grass along a track by the forest and suddenly emerged into a clearing with a beautiful lake. Black herons and a cormorant sat on tree stumps sticking of out of it, a dragonfly was flipping water over its eggs, and we saw a tree full of delicate weaver birds' nests; but what really caught my attention was the pitiful sight of a man sitting by the water who said he worked at a nearby coffee plantation. His legs, ankles, and feet were hideously swollen out of shape, with the skin a horrible crusty-gray from elephantiasis. He begged us for fifty cents and was thrilled when Cliff gave him a shilling, but this was very little help when he had to live with this terrible and, at that time, totally untreatable disease.

On another day we headed off in torrential rain to Murchison Park and discovered there were many fewer animals than when we went through there in 1965. However, this time we got to see the spectacular Murchison Falls and, once the rain eased up, walked over to see the water forcing its way through a gap in the rocks and cascading down for about 140 feet to kick up a huge white spray at the bottom before the river flowed on west into Lake Albert.

Occasionally, we traveled on foot along the main road and stood by the side to let a vehicle go by; and, because we were fairly close to Murchison Park, tourists sometimes passed us in Volkswagen minibuses painted with garish black-and-white zebra stripes. Inside, we glimpsed well-dressed, freshly washed people bedecked with expensive cameras, binoculars, and sunglasses. They looked out in astonishment at the sight of our disheveled little crew with a tall white man and woman accompanied by two small blond-haired children. All the tourists looked well-fed and healthy, but once their vehicles passed in a cloud of red dust, we were back to the Africans in their worn clothing and bare feet, and children with ulcerated places on their legs that were covered with flies and badly in need of medical attention. Men would ride by at a leisurely pace on bicycles with a few fish attached at the back, and the battered old bus made its way slowly between Masindi and Butiaba, breaking down frequently as it went. Then I felt as if we had gone back in time and were living in a dream in which I tried to reconcile these two totally different worlds.

Meanwhile, the baboons were becoming a constant feature in my life, though it had not started out that way.

CHAPTER 36

Surrounded by Baboons

CLIFF WAS NEVER HAPPIER than when he could wander off to observe the birds, animals, plants, and trees in the surrounding area. He explored for hours, taking a notebook and pencil, a bird book and plant book. As long as he could head off to the forest and its environs, he was content and seemed relatively untroubled by the lack of amenities at Busingiro. He accepted life in New York but always wanted to get back to the wilds of Africa.

I lacked his great enthusiasm for the outdoors and liked the fast pace and noise of Manhattan, where I was always busy. Everything came to a screeching halt at Busingiro, which was so different from my expectations. Never could I have anticipated the dramatic change and its impact. We'd been used to living in an apartment building in Greenwich Village, where we had to wait in bare, stuffy halls for elevators. Our windows looked out on concrete sidewalks and tall gray skyscrapers. I had stood on dirty subway stations and been jostled by rush-hour crowds. Millions of people were crammed onto the small island. It also provided a wide variety of entertainment with theaters, concerts, museums, and art galleries.

At Busingiro, I could walk out of the house onto the verandah and the grass interspersed with small bushes that stretched for miles down the escarpment. No buildings obstructed the view, and we saw few people on a regular basis. We were used to endless noise in Manhattan, where car horns blared, police sirens screamed, fire engines clanged as they raced along with lights flashing, cab brakes squealed, and their drivers hurled abuse through open windows. Busingiro was very quiet, and for a start I craved sound. I even felt disoriented by the lack of it, and began to wonder what happens to human brains when we are surrounded either by silence or loud noise for any length of time. My ears took several days to adjust, and then I became aware of different, more subtle sounds, such as the insistent buzzing from the whirr of insect wings, clicking noises from crickets, snapping of twigs in the undergrowth, songs of birds, the rustle of the wind, and the sighing of the grass.

I was used to dry, polluted air in the central heating of New York winters. My nose had been sore, had started to bleed, and my skin had itched without lotion. These problems continued for weeks after the winter, but in the clearer atmosphere at Busingiro they abated. I smelled damp leaves and earthy scents and experienced the early morning dew on the grass. This took me back many years to my childhood in the English countryside, and I realized how much I missed it.

My eyes also had to adjust. For a start I focused on little apart from the expanse of cut vegetation in front and the mango trees around the perimeter of the property, but then I began to appreciate the different shapes of the leaves and the shades of green and brown. I could look for miles across the escarpment and see the shining surface of Lake Albert with the bluish colors of the mountains of the Congo on the far side. With no industry, the air was clear of smoke and haze. We had none of the amenities to which we were accustomed, but life was much healthier.

Still, I grew extremely bored, mostly because I had nothing to read. We

THE ELUSIVE BABOON

had taken some children's books, but to save on space and weight had taken none for me, thinking there would be something at our destination. I searched the house, but all I found was an old copy of *Time* left by previous occupants. I read it from cover to cover several times while the children played. With no reading material, no entertainment, and unable to head off on my own, I found that, on most days, I did little apart from watch the children, cook in primitive conditions, and hang around waiting for Cliff to come back from the forest. With time on my hands, I would look at my watch, thinking an hour had gone by when only fifteen minutes had passed.

I understood why the children fretted and complained of boredom in the beginning. They had been used to television, lots of friends, a large playground, and many toys. Now they only had each other, no TV, and the few toys we brought with us. For several days they moaned until they realized things were not going to change. After that they became remarkably creative in their play, used the possessions they had, and took advantage of their surroundings.

Erik, whose fifth birthday we had recently celebrated, made tracks through the earth with his Dinky cars: tiny cars manufactured from die-cast metal in England. He and Caroline built pretend houses of sticks and elephant grass before they collected grasshoppers for a game they called "Grasshopper City," in which poor little Erik was assigned the impossible task of making sure that nothing escaped. After a while the children took over one of the rooms and called it their hideout. They dragged in a bench to act as a table and made a notice to put against the door on which Caroline wrote *Home* on one side and *Not home* on the other. I thought it very *Winnie the Pooh*. They watched the sweat bees constructing their little wax tunnels near the back porch, and were fascinated by the frogs and toads in the small reservoir at the back. The little toads made them laugh because they imitated Cliff when he made a noise like a "caw," and then their chorus floated across the night air. He also fed the toads with moths by dropping them on the water, and the children were thrilled when the toads swam over, seized the moths with their little hands, and stuffed them fast and furiously into their mouths.

A fine brightly colored lizard about ten inches long sunned itself on a stone at the corner of the house. Its head and half of its tail were bright orange; the rest of its body and the tip of its tail were black. This creature amused us by lifting its head and arms and going up and down like an athlete doing push-ups. Geckos with pale pink flesh-like skin came out onto the verandah to eat insects in the evenings, and once darkness fell, we ate by the light of the hurricane lamp and read to the children or played card games before bedtime.

But I was still bored—and then I discovered the baboons. Caroline first spotted them on the edge of the property, and for about ten days we saw only one or two; then more began to appear, and we realized that a troop circled the house every day, following the line of the mango trees just beyond the perimeter of the garden. Some days when Cliff had gone to the forest, when Marco was not around, and the children and I were alone, the baboons came quite close to forage, and I began to spend time watching them from the shade near the house.

To my surprise, baboon watching was very interesting. I thought there must be something unique about them, because the ancient Egyptians had revered them, but I had always viewed them as rather ugly, smelly creatures. I had seen baboons on our first trip to Uganda but never had the time or opportunity to watch them interacting with one another. A proper scientific study of their behavior required extensive note-taking, following the troop on its daily rounds, and making recordings over a long period of time. I was a novice with a limited amount of time, and my observations were somewhat anthropomorphic, but I derived intense satisfaction from watching and eventually became so absorbed that I was unaware of anything else apart from the children. My journey with the baboons had begun.

CHAPTER 37

Baboons at Play

MARCO WAS WORKING OUT OF SIGHT at the bottom of the drive, Cliff had gone into the forest, and the children were playing quietly indoors when the peace was shattered by shouting, screeching, growling, and huffing—*Huh, huh, huh, huh, huh*. I jumped up, raced to the verandah, and saw a small baboon rush into the open before it shinned up a tree. As soon as I appeared, the noise died down, but I grabbed the binoculars, told the children to stay in the house, went to the side door, and stepped out to see what was happening.

A baboon was crossing the clearing beyond the grove of mango trees. I moved forward on the grass, stopped, but saw no more baboons. I headed back, stepped into the shadow by the door, turned back to look at the clearing, and saw two baboons walking stealthily across in tandem. They turned and stared in the direction of the house like two actors on stage in a variety show, looking at the audience. I stepped out again, saw nothing, returned to the shadow, and once more saw baboons. I realized they were watching me closely and calculating when to cross the clearing. If I remained in the shadow, they would move, so I stayed there and watched as they all crossed over and went on their way.

A few days later, I heard the baboons again, told the children to stay in the house, and moved in the direction of the noise. I saw nothing, but when I crossed the open part of the garden and headed towards the mango grove, a baboon rushed down from a tree ahead of me, while animals crashed through the bushes on my left. Suddenly, a sharp bark made me jump, and a few feet in front of me a large male baboon emerged from the shadow of the trees to block the path. He must have weighed about eighty pounds, and he had huge powerful muscles and sharp hazel-brown eyes that glared at me from under furrowed brows. I hadn't seen him because he blended so well into the background, but he could easily see me as I stood in the sunlight and walked in his direction. I stopped, looked at him, and, as we stood face to face, time stood still.

Big baboon, large canines, he could tear me apart flashed through my mind, but, most important, *I've left the children in the house on their own. I must survive and get back.*

Trusting the baboon would not attack, and gambling it would stay where it was if I didn't advance, I took a deep breath, turned, dashed to the house, and sat there for a while to calm down. The baboon didn't follow, and I realized he was just seeing me off before I got too close.

After a while, I went outside again. All was quiet, and I thought the baboons had gone away. I saw no sign of the big male as I cautiously retraced my steps, but suddenly there was a sharp bark and again he came out of the shadows, staring hard and blocking the path. My mother's words came back: "Curiosity killed the cat." Maybe I was pushing my luck. I retreated to the verandah and stayed. Soon afterwards the male gave a parting bark in my direction, as much as to say, "We're leaving now. Do what you want!" He must have been keeping guard, making sure the others were safe while they crossed the clearing, and telling me to mind my own business or I could be in trouble.

THE ELUSIVE BABOON

Thus I learned not to go too close and discovered the male was not likely to attack unless his territory and troop were threatened.

As time went by, the baboons became used to us and ventured closer to the house. They would sit in the garden, pull up elephant grass, and peel and eat the tender white stems before discarding the rest. Sometimes they played in the deserted summer house or entered the old kitchen at the back after Marco left in the afternoons. As long as we kept at a certain distance, they seemed unafraid, but they were very wary of Africans. The locals viewed baboons as a menace because they raided crops and stole staple foods such as maize. They chased them away by hurling sticks and stones. They also set wire snares to catch antelope (duikers), but these often caught the legs of baboons or chimps and led to a loss of their hands and feet. A baboon in our troop had a piece of wire around one of its paws, which must have come from one of these snares. The wire had cut into the animal's flesh, caused a loss of circulation, and now the paw hung limp and useless.

Our baboons loved mangoes and gorged on them when the trees were in fruit. They held the fruit in one hand, put up the thumb to balance it and guided it to their mouths with the other hand. As they did this their eyes darted around on the alert and their eyelids went up and down showing the white lids like a warning signal. When there was an abundance of fruit, they became wasteful and took only one or two bites before throwing the remainder on the ground. This reminded me of a time when my mother's cousin visited. While the grown-ups were in the dining room drinking tea, the children had gone into the sitting room, where they found a large bowl of apples. My mother later discovered the apples discarded behind the sofa. A bite had been taken out of each one. This was much like the baboons with the mangoes.

When the mango trees had less fruit, the baboons ate a wider variety of foods and fed on the roots of the elephant grass, the leaves, berries, and seeds of certain trees, and enjoyed the fruits of the guava trees that grew in a line by the side of the kitchen. They drank from the reservoirs behind the house, and we saw their brownish-green backs bending forward as they flexed their arms until their mouths reached the water.

Our troop totaled about thirty, with the dominant male distinguished by his size, the huge mane around his neck, and his nose. This had a distinctive split that was probably the site of an old war wound. "Split-Nose," as we called him, generally sat alone while the others kept a respectful distance. The only animal who went near on a regular basis was the female with the smallest baby, who seemed to approach him for protection.

Our troop had another big male in addition to Split-Nose. These two ventured furthest into the garden, although Split-Nose, as the leader, came closer to the house. Often he sat on his haunches, threw back his head, and opened his mouth in a huge yawn to display ferocious-looking canine teeth about one-and-a-half inches long, with dagger-like ends. They could inflict a deep puncture wound or rip flesh apart and were guaranteed to instill fear into anyone who came in close contact. Initially, I thought the yawn meant the animals were resting and tired like humans, but I was wrong: They were actually very alert. If we came within a certain distance, a male was likely to yawn before sending out a warning bark, at which the rest of the troop sitting behind him started to retreat.

Young adult males fought frequently among themselves but obeyed the two big males without question, and if a younger male approached an older one and was not wanted, the latter had only to bark for the younger one to back off. Yet we saw these huge male baboons playing quite gently with very small baboons, and I once saw a female leave a small baby in the care of the second largest male. He watched over it and allowed it to jump on him in a manner he would never have tolerated from youngsters independent of their mothers.

Younger members of the troop formed several small peer groups. One was composed of three sub-adults, and another had three juveniles who constantly fought and played together. Infants, who were not fully independent, hung around the females, who generally stayed with one another, especially those with babies. Females sat in groups and watched the smaller baboons at play like women who sit and talk together when they take their children to the park. When ready to move, they summoned the little ones with a bark, and then the smallest babies, who still had pink faces and Mickey Mouse ears, would each jump underneath its mother, where it could cling to her once the troop set off. In this way they got the most protection when she walked along on all fours. Young baboons, whose faces were turning from pink to black and were unable to travel far on their own, would jump onto their mothers' backs, where they rode like jockeys on horses.

After they became used to us living in the house, the troop heralded their approach around the back with the noise of fighting. This gradually grew louder, and a host of small birds started to twitter and fuss when disturbed by their movements. First came two or three of the young males, with one up a tree to act as a lookout. Next, Split-Nose appeared and chased away the young males. After him came the female with the youngest baby, and gradually the rest of the troop appeared, but in no discernible order.

Generally, they headed to the mango grove, but they also liked to go into the summerhouse and had made a track to it from the grove. We weren't sure why the baboons went there but, in their absence, found holes in the concrete floors. Because we had seen several of them stick their paws into a hole outside the building before they extracted and ate something, we thought they probably did that inside the house as well. On another occasion a baboon emerged from the summerhouse carrying a piece of red brick. He carried it to the edge of the garden, sat down, and chewed on it, causing us to wonder whether some kind of mineral existed in the holes and brick that supplemented their diets.

Fights, accompanied by squealing, barking, and shrieks, often broke out among young males in the summerhouse. Several animals then flew out the windows or through the open door before they scooted onto the roof or shinned up some eucalyptus trees nearby. Occasionally we saw the older males intervene and chase the youngsters away, but other members of the troop sat outside, groomed one another, and took no notice.

In the late afternoons our troop usually headed down the drive to forage, and toward evening went to an abandoned house just off the main road. Its roof was missing, but some wooden joists still remained, and one evening we watched several baboons lying back on them to bask in the sun's rays. Small animals played close to their mothers. Females groomed the males, each other, and their offspring; and one or two animals even groomed themselves. Grooming was important to help clear away ticks and other parasites, in forming physical and social contacts, and keeping pelts in good condition. One small baboon stood on all fours while its mother groomed it meticulously along its back. Then it turned around and poked her playfully, which she placidly accepted before it cuddled up to her and she cradled it, leaned over it, and fondled it around the head like human mothers do with their babies. However, one female never received any grooming and had a neglected, scruffy appearance. She would emit pitifully small, sharp barks and was mostly on her own. We called her "the neurotic baboon," but apart from being neglected and her odd barks, she seemed to eat well, had plenty of energy, and appeared healthy. We never discovered why the others ignored her, but I speculated it might be because she was not fertile and had no children. Overall, the scene was one of relaxation and amity, but as the sun was setting, most of them moved on to prepare for the night. We returned to Busingiro to settle in and plan a trip to the forest.

CHAPTER 38

Butterflies and Baboons

THE NEXT MORNING, WE WERE OFF early with the children chattering excitedly on the back seat of the car. At the bottom of the drive, we turned right onto the main road and rattled along its rough surface of stones and red soil towards the forest.

After about a mile, Cliff swung the wheel and headed up a slight bank among the trees onto a track cleared by the forestry department to take a four-wheel-drive vehicle. We parked at the side and got out. In contrast to the sun's glare on the road, we found ourselves in soothing shade. Dense vegetation of shrubs, lianas, and bushes in restful dark browns and greens grew to the side of the track. Trees soared skyward, some so tall you had to bend over backward to see their tops.

Insects buzzed, birds twittered, and occasionally the sound of water running in small streams or dripping from leaves filled the air. Branches brushed against our skins and sometimes pulled at our clothes as we set off. The rains had been heavy, and a fresh, earthy smell rose up from the red-brown soil and soggy vegetation. Vehicle tires had tamped down the track, but it was still muddy and slippery in places, so we trod carefully.

Cliff slung his binoculars around his neck, put his *Birds of East Africa* and a notebook in his pocket, carried his rucksack to collect baboon foods, and led the way. The children skipped behind while I brought up the rear, somewhat fearful of danger from snakes and insects. Cliff was intent on observing plants, identifying them, and making notes, which fortunately slowed him down, because after we had gone about fifty feet we rounded a corner and came face to face with about twelve baboons. One immediately gave a warning bark, at which I grabbed the children and whispered they must be still. A baby ran to its mother, jumped on her back, and the two of them watched from the side of the forest track. The dominant male moved into the middle of the track, sat down, and stared at us. Every so often he turned away, put his nose in the air, and yawned as though completely disinterested. But we had learned from watching our Busingiro baboons that this was a sign of nervousness and a warning. We needed to keep our distance. Three females came and sat near the big male. A small baby nestled close to one of them.

Suddenly the large male rushed into the forest but came out again almost immediately, followed by a young male, and resumed his place. After watching for a while, we wanted to move on and began to edge forward with the big male eyeing us. He stayed put while the females got up and crossed the track into the forest. But the young male started up the track towards us. At this, the big male let out a sharp bark, which stopped the young scamp, who dove into the bushes. Not until they had all departed did the dominant male leave and allow us to go through. We gathered the remains of their foods, stored them in the rucksack to analyze later, and I thought this was so much easier than trying to trap animals to take blood, as we had done on our first visit.

Treading as quietly as possible, we headed deeper into the forest until suddenly we heard leaves rustling overhead and realized we were being observed. Looking down on us with quizzical eyes were some Colobus monkeys. Magnificent white shaggy manes hung down from each side of their black backs, and their long, shaggy white tails dangled from the branches on which they sat, but most distinctive were their furry black heads, which gave them the appearance of little old men in smoking caps. Some blue monkeys were more difficult to spot because their bluish-gray coats blended so well into the background, but one shouted at us from the fork of a tree. These monkeys emitted three distinct sounds: One was a sharp bark, another sounded like a bird chirping, but most characteristic of them was a third sound like someone sneezing. Some red-tailed monkeys raced along the top branches, then leapt into the air like small red flying machines and hung suspended in space before landing gracefully on another tree.

And then the trees opened up to reveal patches of light and small pools formed in the track where torrential rain had etched hollows and filled them with water. In front of us a scene like an Aladdin's cave full of jewels opened up as butterflies flitted around in a kaleidoscope of colors. We saw tiny ones in clear yellow, velvety mauve, deep violet, orange, fawn, and white; larger ones in jet black, brilliant turquoise, or sky blue with bright patterns; and a white one with wings edged in a pattern like coffee-colored lace. Some had triangular-shaped wings with sword-like edges. Others had softer outlines with B-shaped wings, and some had rings like two eyes on their hind wings. Orange and brown ones landed on dark green leaves. Scarlet ones floated like the petals of a flower. Many of them dove down and flew up before scattering in the air like fragments of a rainbow. Some gently opened and shut their wings when they landed and basked in the sunlight as if drinking in its warmth. We had entered a lepidopterist's paradise.

Our children loved the butterflies and skipped about, trying to catch them in the air or on the ground when they drank from the shallow pools of water. Erik caught one, but for the most part the butterflies were too quick. They stayed quite still on the path, moving only when we were upon them. Once or twice, we got so close we were sure we would catch them, but at the last minute they flew into the air, avoided our outstretched hands, and flitted off, ready to tease again.

After two hours we retraced our steps to the car and drove onto the main road, where we saw a troop of baboons ahead. We stopped, got out, and followed at a distance on foot to watch as they ate grasses and ginger-flavored *Aframomum* plants, plus various berries from an area cleared of trees. The dominant male sat in the middle of the road with his huge bulk in full view, keeping a watchful eye on us while the troop advanced ahead of him until they came to a patch of forest where they dispersed to shin up some *Maesopsis* trees with long thin grayish-green trunks that soared into the air. Towards the top, their branches fanned out like an umbrella to provide shade. They had fruits about the size of an olive that turned from green to black as they ripened. The baboons stuffed their faces with them like children eating Halloween candy and sought them out so avidly that I wondered if they filled some medicinal or specific dietary need. Maybe we would find out, because we added the remains to our collection after they moved on.

We got back in the car and soon came across another baboon troop. This time we pulled up, stayed in the car, and watched. Ahead of us we counted thirty-eight animals. They seemed unfazed by people, and if anyone happened to ride by on a bicycle, or if the occasional car went by, they ambled to the roadside at the last minute. From there, they peered out from tall grasses, waited for the person or vehicle to go past, and then strolled back into the road, where they sat down in small groups. Among them were four mothers with babies. One mother pushed her baby away playfully; then it leaped back toward her, and she cuddled it. Other babies left their mothers for a while but returned as if seeking comfort and reassurance. Some little ones ran up to their peers, and they pushed at one another before rolling over and turning somersaults. One leaped into a steep bank at the side of the road, ricocheted off, got up, and gamboled off.

The males, as usual, were much more active than the females, who generally sat in groups with their young, groomed one another, or groomed a

THE ELUSIVE BABOON

male. On this occasion, we saw a large male lying flat on his stomach in the middle of the road with his head to one side while a female groomed him. Like a customer at a massage parlor, his face had a look of unadulterated bliss. He stayed until a large car came along and then, with extreme reluctance, got up and shifted to the road-side.

Three big males paraded around this troop. Suddenly one of them went up to the dominant male as if to challenge his leadership, and we expected trouble. But the dominant one was in prime condition and soon dealt with the usurper by chasing him into the bushes.

As in other troops, the members of this one were sociable. Many sat in small groups enjoying the morning sun as they ate and played a little. They seemed to communicate with facial expressions, posture, touch, and low sounds like grunts. But their communication changed dramatically to hoots and barks when they were in the trees and undergrowth, where they were unable to see one another, or if they felt threatened.

When I saw how much the animals interacted and how much the little ones depended on their mothers, my mind wandered back to Nuts, a baby baboon who came to us in 1966, just after the political crisis when the Kabaka fled the country. A small African boy brought him to Cliff, saying the baboon had been a pet of the Kabaka: "He royal monkey, sir." Cliff put the small creature in a large cage he'd set up in the garden. It contained a swing and a sleeping area. He thought, if he took Nuts to the field, Nuts would lure other baboons into the cage. Then he could trap them and take blood. It never worked. Nuts became very nervous, and the other baboons kept their distance.

Nuts loved bananas, which he stuffed down at great speed with his quick eyes darting from side to side. In this way he was like the baboons near Busingiro, but in other ways he was completely different. His name came from his strange behavior. When people appeared, he carried on in the most ridiculous fashion, and we always knew if someone was approaching because he screeched loudly and leaped around bashing the sides and top of the cage. One day I looked out the window when the noise started and saw an audience of Africans sitting on the grass. Nuts was going wild doing double somersaults all around the cage before he rushed to the top, hung upside-down, swung from one arm, and fell off. After this he pretended to fall off a stick set up in the middle of his home but at the last minute grabbed a bit of wire projecting from the top of the cage and hung from it suspended by one arm. Following this he ran to the cage door and banged his head on it several times.

His audience laughed and applauded. The more people laughed, the more excited Nuts became, and he ran around baring his teeth as though grinning at everyone until the audience was in an uproar, clinging to one another and howling with mirth. We knew nothing of his past and what provoked this behavior, but he was much quieter when we went to see him. He might execute one or two somersaults, but then stopped, walked around, or went to lie in the area that became his bedroom. Caroline's visits seemed to be the highlight of his day. She would go outside, talk quietly to him, and give him wild raspberries from our garden, which he loved.

Nuts could be amusing, but I also thought he was a sad little creature, especially after I watched him once or twice from our doorway when no one was around. He sat on his own, biting his nails, or he sat with his legs crossed and swayed back and forth in a schizophrenic fashion, reminding me of films showing babies abandoned in institutions with very little human contact. He was quite young, and it was sad to see how much he clung to Cliff when he held him. When I saw the baboons near Budongo living in troops with thirty to forty members, and how the little ones clung to their mothers and played with their peers, I realized Nuts must have been very lonely. But he also had little in common with wild baboons. To return him to the wild would have been almost impossible. He had not learned to defend himself, had not learned how to forage for food, and had no troop to which he belonged. If put near a troop, he would probably have been attacked as an outsider. But he also had no fear of humans and could be a menace as he grew older and stronger. We

THE ELUSIVE BABOON

kept him for four months, until it was time to leave Uganda, at which time some people on the agricultural station agreed to take him. The last we heard he was once more proving to be the star turn who entertained the neighborhood. But after our experience with Nuts, I strongly believed people should never keep wild apes and monkeys as pets.

I came out of my reverie and realized it was time to go. We drove forward, and the baboons moved to the roadside. Their eyes looked out from the bushes and grass as they waited to resume their positions on the road when we went by. We'd had a successful morning collecting samples, seeing butterflies and baboons, and had discovered, if we stayed on the forest track and kept our eyes and ears open, we were safe. But our quiet routine was about to change. Some visitors were due to arrive at Busingiro House.

CHAPTER 39

Culture Clash

FOR SEVERAL WEEKS WE WERE JOINED by two female American students who wanted to study baboon and chimpanzee behavior. One of the women had dark brown hair; the other was very fair. Like other nationalities that travel abroad, Americans have their own distinct peculiarities. The Americans who came to Budongo arrived with cases full of medicines and deodorant sprays. We thought they had an exaggerated fear of germs and bacteria, and we found their strong reluctance to eat the local food odd. This was the first time that either of them had spent time in the field, and for a while they were traumatized by the whole scene, especially the bats living in the roof. Meanwhile, the air was filled with their hair and deodorant sprays in the mornings. Most amazing to us was the vast amount of toilet paper they consumed. Neither of us had the nerve to ask what they did with it, but we found ourselves conspiring to grab as much toilet paper as possible on our visits to the hotel in Masindi when they insisted they needed more. No doubt they thought we Brits were very peculiar to put up with the conditions at Busingiro house, and I understood that.

THE ELUSIVE BABOON

People from other countries can also be amazed by the way Americans dress. We experienced this one afternoon when one of the students came back in fits of laughter. She said she had been exploring the area near the house in her "afternoon baboon outfit," consisting of brown boots, blue jeans, and a white T-shirt covered at the front with the round yellow balloon shapes of "happy faces." Each face had two black eyes like round raisins, no discernible nose, and a black line in the shape of an upturned smiling mouth. She had topped all of this with a dark-green plastic eye shade to keep the sun's rays at bay and avoid skin damage.

As she wandered through the woods in this "normal" garb, she heard a baby crying, moved towards the sound, and in a clearing came upon three African women in long dresses and with scarves tied on their heads. They sat on the ground with their washing in a basket and a tiny baby they were trying to pacify. The women were absorbed in their activities when suddenly the American apparition appeared from out of the bushes with grinning yellow faces all over its chest. What the women thought was anybody's guess, but the creature was too much for them. They screamed, leapt into the air, scattered washing all over, dropped the baby, scooped it up, grabbed the washing, and took off down the hill in a mass of skirts. No one would have batted an eyelid in Greenwich Village, but to the Africans near the Budongo Forest it must have seemed as though an alien had descended from outer space.

Our visitors had a great fear of insects. They screamed and flapped at them, and at times went to amusing extremes, as happened one evening when we took the car and went by the usual track into the forest. It was very dark, and we could see nothing, although we heard a lot of different kinds of frogs croaking and making a din before a rock hyrax screamed. This alarmed the two women, and they insisted on returning to the house. Under pressure we set off back. Suddenly, one of them uttered a piercing scream. "Stop the car, Professor Jolly! There's a thing as big as a baseball three inches from your head!"

Cliff slammed on the brakes, whereupon any altruism that might have existed in the women was abandoned. They piled out of the back of the vehicle, pulling poor little Erik with them, and plunged the rest of us into the darkness with the "thing," because they had taken the flashlights with them. After a minute or two of cursing, Cliff lit a match only to find that the "thing" was a spider. He calmly knocked it out of the window, the fleeing masses climbed back in, and we drove on. Our visitors said they were totally shattered by the incident and, somewhat to our amusement, declared that they would have to take tranquilizers when they returned to home base.

They continued to be very unhappy with the way of life and kept gasping for cold drinks, iced beer, Coke, and sodas, but with no electricity and no refrigerator, it was impossible to have anything like that in the house.

Nevertheless, as time went by, we settled into a routine. The mornings were given over to field work. We had a late lunch at about two in the afternoon, rested a while, and wrote up notes. The hours between four and six were used for baboon tracking, and during that time I prepared the evening meal.

Cooking was a challenge. The most economical way to cook was to build a fire in the "kitchen," as this not only conserved our supply of bottled gas but allowed us to cook with several pans at a time on the fire. The main problem was to get the fire started. Patience, skill, and an understanding of which wood to use were necessary. You had to know how to stack the wood, blow on it to get the flames leaping, and then wait until they settled to give a nice glow so that cooking could proceed. With practice, Cliff learned to select the best wood and reduced the time needed to start the fire from about thirty minutes to ten. The main drawback was the lack of proper ventilation in the kitchen, with no duct to lead the smoke fumes outside. If the wind happened to blow in the wrong direction, the room filled with smoke and our eyes stung, watered, and turned red.

With no refrigerator and little opportunity to shop, our diet was limited. In the morning we generally had bread or porridge, which we ate with sugar

THE ELUSIVE BABOON

and reconstituted milk. Lunch was generally canned soup and peanut butter sandwiches. Our evening meal consisted of rice, potatoes, or spaghetti, with some form of canned meat, and a green vegetable and tomatoes. Tuna fish risotto, which could be cooked in just one pan, turned out to be a favorite. All of this was followed by fruit, usually very small sweet bananas about two to three inches long that tasted like strawberries, and instant coffee. Then we faced the task of washing up in bowls into which we put water heated in buckets on the fire.

We also worked out a routine for bathing after we discovered a wooden plug for the bath and began to heat water in buckets and carry them in. This meant we could bathe in a civilized fashion in warm water. But lugging buckets of water was hard, and we learned to do with infrequent baths and to wash all over instead. The children were a problem because they seemed to get covered in mud most days. We solved this by telling them they could take a shower out in the open when it rained, and that worked. Although it was supposed to be the drier season, we had heavy rain with thunderstorms nearly every afternoon. Neither of them seemed to mind when we told them to take off their clothes, soaped them all over, and sent them outside, where they danced around squeaking and laughing in the downpour as the soap was washed off their bodies. Caroline cried and hated occasions when they got soaked to the skin *in* their clothes, but Erik remained unfazed and trotted along, cheerfully singing, "Raindrops keep fallin' on my head."

About once a week, we drove into Masindi to buy food, get fuel for the car, send and pick up the mail, and get up to date on the latest political news. Although Masindi counted as a town, it was very small. One main street ran through it, and there were only three or four additional streets. However, as we knew from out previous visit, it had a hotel for guests. We sometimes went there for a meal and to steal more of the constantly disappearing toilet paper. Most of our provisions, including tinned meat, jams, margarine, porridge, peanut butter, and soap, came from a small grocery shop run by an Indian.

Bread came from a so-called hotel that seemed to be the equivalent of a café, but bread was difficult to keep in that warm, moist climate, and it soon grew a whiskery green mold. By the end of the week we had to throw it away or cut off as much mold as possible and eat the rest. Otherwise, we were happy to eat porridge.

Our vegetables and fruit came from the open-air market, which was divided into two sections. One was devoted to selling second-hand clothing and brightly colored cotton fabrics. The other was enclosed by a wall and devoted to foods. In this section, vegetables were arranged in small piles; for instance, there would be twelve tomatoes or potatoes in a little heap. As in Kampala, women sat by them and tried to sell them. There were also bunches of *matoke*, piles of dried beans, maize, onions, peas, and cabbage. We bought sticks or hands of small sweet bananas, and obtained sweet green oranges, good pineapples, and paw-paw. Meat was hacked up in two small huts at the end of the market where cows' hooves, horns, and entrails were scattered all over the floor. To us this was unhygienic and unappetizing. Yet people crowded around and haggled for pieces of meat as they were cut off. I watched one woman leaving with a cow's leg slung over her shoulder. The hoof and skin were still attached.

Several sections sold fish. Fresh Nile perch, often weighing well over a hundred pounds, was cut into chunks on the grass and sold at a shilling apiece. The flies hovered, ready to descend. Once or twice we bought some very fresh fish, cooked it, and thought it tasted good and made a pleasant change, but our American friends refused to eat it and were horrified at the thought. Two stalls sold dried fish. This was shriveled and yellowish-brown in color, and looked most unappetizing as it lay in the sun surrounded by flies. Meanwhile, the ubiquitous vultures hovered hungrily on the perimeter, ready to swoop down at the first opportunity for their share.

On a number of occasions we had to shop in the market after torrential rain. When this happened, the place turned into a muddy quagmire that we

THE ELUSIVE BABOON

had to plough through on foot, slipping and sliding as we went. On one occasion, the Americans were with us. We were in the car when it stopped and became bogged down on the road as we were leaving Masindi following a particularly heavy storm. Everyone piled out except Cliff who insisted he had to stay inside at the wheel to guide the vehicle while the rest of us pushed it out of the mud. No one else could see why he was more qualified to do this, but as he was literally in the driver's seat, it was impossible to get him out short of physically attacking him. In spite of appeals, veiled insults, and threats like "What gives you the right to keep your ass in there?" and "You're heavier than everyone else—let someone lighter take a turn and relieve the load, so it's easier to push," he hung on, ignored the comments, and instructed us to get pushing. He apparently chose not to see himself as an English gentleman but as the *bwana* in charge of a bunch of women, and was determined not to lose face in front of some Africans who were starting to gather round to see what was going on.

After some argument, the children were allowed to stay inside on the back seat, and the three of us began to push. It was hard to get a grip, and we kept sliding backwards, but after a few "heave-ho's," the car began to shift, the engine sprang to life, the driver gunned it hard, the car shot forward, and at the sudden release we all fell full length into the mud. At this, the African onlookers, who until that point had been looking with mouths open in astonishment at three white women pushing a vehicle, began to shriek with laughter. We picked ourselves up with as much dignity as possible, tried to wipe off some of the thick mud, cursed the driver, who also was enjoying the joke, and got into the vehicle. The last thing we saw was a crowd of folks holding onto their sides and laughing their heads off. Our children watched all of this and tittered in the back. Not surprisingly, they got yelled at as the tempers of the mud-bespattered were, understandably, frayed. That day, we were not a congenial group and were all pleased to return to clean up in some semblance of normalcy at Busingiro. There was more to come.

JENNIFER JOLLY

CHAPTER 40

A Shot in the Dark

BUSINGIRO'S ISOLATION WORRIED ME. Our nearest European neighbor, a Scot, lived fourteen miles down the road, and with no phone or radio to communicate with the outside world, it was difficult to find out on a regular basis what was happening in Kampala and whether we might be in any danger at Busingiro. Our main source of news came from Masindi, but we went there only once a week.

We saw few people. Marco came every morning and our other helper, Yesero, came three mornings a week. Each day, a band of men dressed in shorts and tattered shirts, and carrying scythes over their shoulders, trudged along the track going up the hill through the elephant grass at the side of the property to the sugar-cane estate owned by an Indian on the other side. They appeared at about six in the morning and returned at around four in the afternoon. Women with babies on their backs occasionally came through in groups, chattering and giggling when we waved to them, but I only remember one occasion when we interacted with another African at the house.

This happened soon after we arrived, when a man clad in old khaki shorts and a faded, ragged, short-sleeved gray shirt appeared at the open French windows. He held a stick of bananas in one hand and carried a ferocious looking *panga* in the other. It was the sort of weapon used to slash down thick vegetation and could cause grievous bodily harm. On seeing it, I approached him in trepidation, and as I drew closer could see his eyes were bloodshot. He smelled strongly of alcohol and sweat. When I appeared, he demanded money and new clothes in exchange for the bananas. Too afraid to tell him to go away, I gave him some money, thinking that would be the end of it. But this only encouraged him. He kept coming back and after a while became a nuisance. On one occasion, he turned up to wheedle money out of me and, after I paid him for some tomatoes and maize, he said there had been an accident at the end of our drive. It sounded as though blood, gore, and bodies were strewn all over the road. Cliff happened to be at the house and went to investigate while I waited to hear the worst.

The accident turned out to be much less serious than we imagined. Four Africans, a male driver together with three women in long dresses, sat on the roadside close to the entrance to our drive. One of the women had a large gash in her chin, which was bleeding profusely. The others were trying to stem the flow. The driver said he had fallen asleep at the wheel of his little Peugeot truck, causing it to swerve into the ditch. The women, who were traveling in the open back, had been thrown out. Apart from the woman with the gash, the others seemed shaken but unhurt.

No one knew what to do about the wounded one. After some discussion, we agreed to take her to the hospital in Masindi. We found a cloth to cover the gash, told her to press it tight, and then we all piled in the car and set off. We dropped her at the hospital, shopped in the market, and picked her up later to take her back. She had had stiches in her chin and a shot against tetanus.

As for the man with the *panga*, he continued to come until I decided he would pay more attention to a white man than to a woman and insisted that Cliff put his foot down. After that, no one else came to bother us during the day, and at night we were completely alone.

Our isolation struck home one evening when our American visitors were with us. They were not having a good time. Although satisfied about seeing plenty of baboons, they hated the discomfort of the house, had been frightened in the forest, disliked shopping in Masindi, and from their comments it was obvious that we and our children greatly irritated them. The next incident was the final straw for one of them.

We had been at Busingiro for just over three weeks when we ran into our friend Jonathan Kingdon, whom we knew from our first visit to Uganda. Jonathan was an exceptionally talented artist who had studied fine arts at Oxford and was a lecturer in Fine Art at Makerere University. A tall man with a large smile, blue eyes, and curly brown hair, he exuded energy. Having grown up in Tanzania, he spoke Swahili fluently and, in addition to art, had a passion for wildlife. Like Cliff, he had an intense interest in primate biology and evolution.

Now, five years later, he visited us at Busingiro and took us into the Budongo Forest with our children and three of his five. During this trip, he and Cliff decided to go into the forest with one of the foresters to see animals, such as bush-babies and pottos, that came to life after dark. It was arranged. To my surprise, the student who had blond hair wanted to go along.

On the appointed evening, Jonathan arrived in his Land Rover, and the three of them set off. The sound of the engine faded as they went down the driveway and disappeared around the curve towards the main road, and I was left with the children and the other American. For a while, the children chattered about the baboons and we played a memory card game, but once they were in bed an uncomfortable silence descended. The main noise came from the subdued buzz of insects and the occasional caw of the toads at the back. Our Calor Gas lamps had given up, and we were a bit edgy because it was gloomy in the dim light of the flickering, sputtering hurricane lamp. Outside it was almost pitch black, with hardly any light from the moon. The evening stretched ahead.

Before long the student became nervous, said she didn't like being there, and found it creepy to be so cut off from people. She started to talk about the stories we had heard of bandits in and around Kampala. After that she began to worry about rumors we had heard that the Ugandan army was out on the roads. She wondered why it was there and whether it was dangerous for us. I said she shouldn't worry; it was not strange to see the army, because there was a base in Masindi. I was afraid of being trapped in the web of her fear and tried to change the subject, but her mind was set on problems. She flipped back her long dark hair and said, "I'm glad we didn't go to the forest! It's bad enough during the day, and I was scared when we went in the evening. I'd be terrified at night."

I felt sorry for her. Neither of them realized what they were letting themselves in for when they decided to join us to watch baboons. I could understand that they were not happy with the facilities at the house but thought they might look on their time there as an adventure and accept it for a few weeks. Besides, Cliff and I had grown up in austerity Britain after World War II, when we had to make do with what we had. Maybe this made it easier for us to adapt to life at Busingiro. In any case, Cliff never seemed troubled by hardships in the field. But the Americans had grown up in prosperity, surrounded by abundance and comfort, so they resented their absence.

While we waited, I thought about the time when the blond-haired student had complained because we had nothing with which to defend ourselves. Coming from the gun-toting South of the States, she thought we should have a gun. She had grown up with them. "Guns were a way of life, ma'am!"

I came from a country where guns were not carried. When I first arrived in the States, I was unnerved when I saw cops in Washington Square Park with guns slung on their hips. They seemed a far cry from the friendly British Bobby. In my view, guns were dangerous and could do more harm than good, especially if they got in the wrong hands. She didn't buy this and said her daddy had taught her how to use one. Maybe she was a good shot, because daddy was no cowboy but a top-ranking officer in the U.S armed forces. She wished she had a gun at Businigiro. I said I was glad she didn't, because she might shoot someone by mistake.

She tossed her head and sniffed. I knew we would never agree, but that night I was frightened to think we had no means of protection. I was pondering on this when the brunette started up again, saying she would be glad to get back to the States because she was scared all the time. As she went on, I realized the one with fair hair was much the tougher of the two. She was a few years older, had experience of the world, and was very smart and not afraid to speak her mind. The brunette was just a college kid. I saw her face start to crumple with fear. "How long have they been gone? Why are they taking so long?"

I thought she was going to cry and tried to reassure her that they'd been gone for about three hours, would be back soon, and we would be fine. But I too was getting more anxious in that isolated place and wished the others would return. It seemed ages since they left. I was relieved to hear the sound of a vehicle entering the drive at around 10:30.

But the vehicle stopped, and the engine cut out before it came into view. The student rose and looked out the French windows. "There's no vehicle. What's happened?"

"I don't know."

"Do you think it's them? Could it be someone else?" she asked.

"I can't tell, because I can't see the vehicle. It's stopped before the bend in the driveway."

Suddenly it occurred to me that it might *not* be them, and adrenaline coursed through my veins. What if it was an attacker from the army? The student had taken her flashlight to the window and was sending out a beam of light to greet them but then said in an anguished whisper, "I can see someone moving out there. Oh, my God, someone's carrying a flashlight up the hill.

They're coming towards the house."

"Are you sure?"

"Someone's out there, I tell you. They're coming this way." She jumped back into the room. And as she did so, a loud shot rang out in the darkness. She squeaked and gasped, "Oh, my God, Mrs. Jolly. What shall we do?"

I also was shocked and unnerved by this turn of events, but my overriding concern was for the children. I had to protect them and needed to make a decision.

"Turn off the lamp and lie down. You're a sitting target where you are."

We threw ourselves onto the wooden floor of the living room, where the earthy dust tickled our nostrils as we lay flat on our stomachs. I was terrified I would sneeze, and that someone would hear. If we were quiet and the place looked empty, maybe they would go away, but then I remembered to my horror that our car was in the driveway. Meanwhile I could hear her moaning, "Oh, my God, oh, my God."

"Shut up and lie flat!" I snapped. "I have to make sure the children are safe."

I lifted myself carefully and crawled through to the bedroom, where I was relieved to find them out of range of the window and the shutters closed, though that was small comfort against attackers. Then I crept around to the side door, which had been left open for Cliff to come in, and locked it. I crept back to the room with the shutters, where the student joined me. I could feel myself shaking. The children stirred, and I begged them to be quiet. We waited with bated breath. To her credit the student, remained still. Then we heard an engine starting up.

"Please let it be them," she wailed.

"Amen," I muttered.

After a few minutes, a light appeared through a chink in the shutters, and we heard a vehicle draw up at the side of the house. Muted voices followed, a car door slammed, and then someone rattled the side door before a familiar voice shouted, "Why the heck is the door locked?"

I lifted my trembling limbs from the floor and opened the door. Cliff and the student came in as Jonathan drove off.

Cliff stared at me. "What's going on? You look white as a sheet."

I was still shaken and could hardly get the words out. "We heard someone shooting in the garden. We were terrified."

"Oh," he said after a pause, "that was nothing. Jonathan saw a rabbit, stopped the Rover, got out and took a shot at it."

I gaped at him. "He did what?"

"He took a shot at a jack rabbit. Didn't get it, though."

"Rabbit! But what about *us?*"

He frowned looking genuinely puzzled. "I don't know. What about you?"

"We saw someone creeping up the garden in the dark with a flashlight, and when we looked out there was a shot. We thought they were coming to shoot us."

He laughed in amazement. "Why would someone come and shoot *you?* Anyway it seems highly unlikely that someone would come here on the very evening Jonathan and I were going to the forest just to shoot you!"

"But that's the point. They knew you were gone, and that we were alone and defenseless. Maybe they wanted to steal something."

I was being illogical but, like the student, was confused and upset.

Cliff had been having a good time and was in no mood to deal with hysterical women. He tried to be reasonable. "I'm sorry you were frightened. But think about it. You knew we were out. All we did was stop on the drive and wait while Jonathan got out and shot at a rabbit. End of story."

"How were we to know Jonathan was shooting at a damned rabbit?"

I began to feel foolish and cross. Meanwhile, the student gathered herself from the floor, looking frightened and furious. She dusted herself off, joined her colleague, and the two of them retreated to their room.

In retrospect, I could see the funny side of it. But six days later, when we

THE ELUSIVE BABOON

saw our Scottish neighbor, he told us that a man named Don Bagley, who owned a small guest house and contracted out labor, had been shot at when someone put a gun through his window. Apparently people seen as having some wealth were not safe and likely to be robbed. Suddenly the rabbit incident no longer seemed so ridiculous and amusing. We already knew Kampala was unsafe, and we had heard the army was out on the roads but weren't sure why. And then the problem with Yesero occurred.

CHAPTER 41

Yesoro's Tale

ALTHOUGH THERE HAD BEEN NO ATTACKS at Busingiro, my anxiety increased after we heard someone had been shot at in their home. And then there was another incident involving Yesero our house-boy.

Marco had turned up soon after we arrived with Yesero (the "s" pronounced as a "th"), saying the boy would clean the house and do some washing, and that he very much wanted to earn money. From our previous experience in Uganda, we knew the Africans looked to white people to assist them financially by providing work. There wasn't much to do at Busingiro, but we agreed to hire Yesero for three days a week on a part-time basis.

Yesero looked about fourteen. He was small and skinny, with large brown eyes and a ready smile. He was pleasant to have around the house and reasonably efficient. He spoke a little English, having learned it in school, and as a result, we communicated quite well. His main job was to wash the clothes by hand using cold water and bars of soap. He would struggle with buckets of water from outside, go to the bath in the spare room, slop the water over the edge of the tub, bend over, and pummel a few clothes around while rubbing them with the soap to remove any grime. Then he squeezed out the soapy water, rinsed the clothes, and hung them to dry over a line of string we fixed up in the back. In this way we kept reasonably clean. Needless to say, nothing was ever ironed, but nothing got terribly crumpled either, and as T-shirts and jeans were our basic attire, and an un-ironed effect was quite fashionable, we were well satisfied. He also fetched some firewood for cooking, swept the floors, and washed a few dishes. He was supposed to work from nine in the morning until one in the afternoon but generally took much less time, and once he had finished his tasks, I let him go. He was pleased about this and even more pleased with the money he earned.

We never found out where Yesero lived, although we knew there were villages hidden down the escarpment. He and Marco probably lived in the same village, but each morning he came on his own, was reliable and cheerful, and carried out the duties asked of him.

One day fairly early in our stay, he asked if he could go home early because the army was out on the Masindi road shooting and killing people who were not working and were not paying their poll taxes: He feared he would run into trouble. Marco corroborated his story. They said they would listen to the radio to see if it was safe to come the following day.

We too had seen army vehicles on the roads but, knowing there was an army base in Masindi, didn't find their presence too unusual. But when Yesero came to me, I was alarmed to hear there might be problems. Cliff was out on the roads, and I had visions of him running into trouble or seeing dead bodies, but he came back and had seen nothing.

The following day Marco and Yesero appeared and reported that, although the army was not killing people, they were in trouble if their taxes were not paid. They never explained what they meant by "trouble."

Time passed until one day, towards the end of our stay, Yesero didn't turn up. Ten o'clock came and there was no Yesero. At eleven there was still no sign of him. With no phone, no radio, and no one to ask for news, I waited and wondered and, as my anxiety increased, began to pace on the verandah. The children played quietly indoors. There was little sound apart from the usual buzz of insects and occasional noise from the gray plantain-eaters that flew across the garden emitting a sound like old ladies laughing at a rude joke at a tea party: *Waaah-ha-ha-ha-ha-ha-ha-ha.*

At around noon, when the sun reached its zenith and I had almost given up hope, Yesero's boy-like figure appeared around the corner of the driveway. He was trudging along slowly, limping slightly; his feet were bare and his shoulders hunched. His curly black hair looked wild, and his clothes were in disarray. Instead of his usual smile, he was trying to fight back tears.

I was very glad to see him but distressed by his appearance and took him inside, gave him a cup of water, and waited. Gradually he calmed down and, struggling with his English, told me what had happened. The gist of the story was that he had been getting ready to come to work when members of General Idi Amin's army came into his village. He said they were scouring the countryside and beating up, even killing, men who were not wearing shirts. People had donned tattered, badly torn garments, to try to stem the brutal onslaught. Yesero had run into the bushes when they reached his village. He had hidden there with some other villagers until they heard the commotion die down and felt safe to come out. Then he'd made his way to us.

I stared at him. Could this be true? Was it possible that people were being attacked and even killed because they were not wearing *shirts?* But why would he make up such a strange story? It was hard to believe, but there was no doubt that something had upset him. I began to wonder whether the army was coming up with some kind of pretext for random brutality; if so, the news was extremely disturbing.

Although we didn't know where Yesero lived, we knew he walked to work, and, presumably, his village wasn't too far away. My mind filled with questions. Did this mean the army was close to Busingiro and they might turn up? Exactly how close were they? What were they doing, and why? We had heard of trouble in Kampala, rumors that people down the road were threatened with guns, and now we were faced with this strange story from Yesero. We didn't know what to think. On the surface, things returned to normal and there were no more incidents, but I was afraid. We made sure the children stayed close and began to put our belongings together. We had been there for just over five weeks, long enough to collect many food samples. It was time to get out.

On the morning of our departure, Marco and Yesero helped us pack the car. We gave them small gifts, said goodbye, and felt sad when we left and saw these loyal young men waving farewell from the house as we went down the winding drive for the last time. We turned onto the familiar red *murrum* road and kept going steadily until we reached the outskirts of Bombo, about twenty miles from Kampala. There we saw banana leaves and branches scattered along the roadside. A crowd of people had congregated near an arch on which words were inscribed. A bark cloth was draped over it. Cliff said the banana leaves and bark cloth were signs of the Kabaka, but we were mystified because the Kabaka had died in London. However, the banana leaves continued, and at each small town we came to, there was an archway with greetings and bark cloth.

Finally we came to an arch where a screaming, cheering, mob surged forward, causing us to stop. From the other direction, a police car appeared. Following it was a large black car which was besieged by the crowd. A man at the side of the road told us the Kabaka's son was in it. He was over on holiday from school in England, and people from all around had been preparing to greet him for days. The enthusiasm displayed for this sixteen-year-old boy was a sad reminder of how his father had been so revered, and the position he had held before he was driven out in 1966, fled to London, became a social worker, and died in terrible poverty. Now his son had been allowed back. The banana leaves continued all the way into Kampala.

JENNIFER JOLLY

CHAPTER 42

The Queen Elizabeth Park

IN KAMPALA WE WERE SHOCKED to learn the area around Masindi and Budongo had been declared unsafe two weeks previously. We had been living there unaware of this, but the news confirmed that our fears had been well founded and that Yesero's strange story may have been true. Meanwhile, numerous rumors were spreading in Kampala about thefts and people being attacked both in and around the capital. Things were worse since the overthrow of Obote because his special security guards had become brigands who roamed around terrorizing people. Armed and dangerous, no one knew of their exact location. Stories circulated of people being held up at the point of a machine-gun and robbed. One morning we heard that a car had been stopped on a road outside of Kampala and one of its occupants killed for no apparent reason. Such cases were frequently reported by word of mouth or in the local *Uganda Argus*.

We heard hostility expressed towards the Indian community, with persistent rumors that the Indians were going to be thrown out of the country. Americans, Canadians, and Europeans were said to be safe, but that wouldn't help if we were caught in cross-fire or in the wrong place at the wrong time. Rumors of political maneuvering flew around. One claimed there was internal fighting in the Ugandan army. Another said the army was fighting with the Tanzanian army and Obote on the border between the two countries. Both could have been true. We were horrified when told Amin's army had been going around villages and killing people. This supported the stories of Marco and Yesero. Men had been seen dead on the Gayaza road leading to Katalemwa estate, where we had lived on our previous visit. Many expatriates were eager to get out of the country, but others said there was not much to worry about. Tourists were still arriving, making it hard to know what to believe.

Ten days were left before our plane departed from Nairobi, and we had planned to spend much of that time in Kenya. We expected to spend two days in Kampala to have the car over-hauled, and to have the baboon foods fully dehydrated to prevent them from rotting and make their transportation easier. To do this, they were put into a kiln in the biology department at Makerere, but an unanticipated setback occurred. Instead of the expected one or two days to dehydrate the food, we were told it would take four or five. We had time on our hands.

By then we didn't feel safe in Kampala and knew it would be hard to keep the children cooped up for several days, so I persuaded Cliff we should go to the Queen Elizabeth Park in the southwest, stay for a day or two, and come back to pick up the dried food. I had never visited that part of the country, the area was said to be safe, and the Nuffield Foundation of England had a lodge for visiting scientists where we could stay. The park was about three hundred miles away on a route that was fairly free of traffic, and since Cliff's last visit to Mbarara in search of baboons in 1966, a tarmac road now covered most of the distance. General Amin was due to open a section just being finished, but the rest was in good shape. People said you could drive there in five to six hours in a reliable vehicle. Our car was slow but we could make the journey in a day. Therefore, we took the car to be serviced and overhauled, bought maps of the area, stocked up on provisions, packed, and set off early one morning. Had we known in advance what we would face, we might not have gone.

The trip got off to a bad start when, thinking we had taken the wrong road out of town, we turned back into Kampala to start again. By then the morning rush was underway and we saw an accident at a roundabout. We chugged along and spotted an African in a brand-new Volkswagen Beetle coming out of a side road with *STOP* written across the road surface where it joined the main road. But the man ignored the sign, continued on, and headed straight towards us. Cliff leaned on the horn, but the man kept moving forward, at which point Cliff accelerated, but it was too late. The Beetle finally came to a halt when it hit us on the rear wing. A moment earlier, it would have crashed into the door next to Erik. Upset and angry, we stopped and got out. The other driver also got out and began to wring his hands in despair. It turned out he was a new driver who, to our astonishment and disbelief, said he didn't know how to stop the vehicle. Fortunately, our car sustained little damage, but his had caught in our rear bumper, which had ripped out his headlights and torn the front. Once we knew Erik was safe, we felt sorry for the driver, but he was obviously to blame and knew it. As usual, a large crowd gathered. People argued and waved their arms until we resolved the situation and left. But by then we were behind schedule.

We saw three more accidents before we cleared Kampala and headed southwest to Masaka, about eighty miles away. Having passed through an intensely cultivated and densely populated area, we drove through an arid plain of scrub with some areas of swamp, seeing few people. The only place of note was a small settlement at Lyantonde, on the border of Ankole province. Forty-four miles from Kampala, we marked our passage over the equator with a bang when a tire went flat and we had to jack up the vehicle, loosen the wheel nuts from their rusted bases with cooking oil, remove them with the wheel wrench, get out the spare, and change the wheel. It took over an hour. At a roadside repair shop we had the tire mended, changed the wheel, and put the old one

THE ELUSIVE BABOON

back. Then we tried the key in the ignition; nothing happened. The starter motor had gone, and the car had to be pushed to turn the engine over. Apparently the car had not been properly serviced in Kampala, and we wondered what other trouble lay ahead.

On reaching Masaka, we headed due west towards Mbarara about eighty more miles, but the tire went flat again ten miles before we got there. Knowing we had no ignition, we managed to stop on a hill, where we propped up the car with stones and changed the wheel once more. We rolled forward, and the engine kicked over. By then it was mid-day. We were hot, getting covered with black grease and red dust, and regretting our decision to make the journey, but continued on as we were halfway to our destination by then. We would get the car fixed along the way.

At Mbarara we stopped at a garage to buy a new inner tube, because the one on the "good" tire seemed to be rotten and the one on the spare not much better. We learned that the starter motor had not been fixed, and that the garage had no inner tube. This unexpected information unsettled us, but again we decided to press on. We had the old inner tube repaired, filled the tank with petrol, and continued west into a more fertile region for about thirty-five miles. Then we began to climb gradually before we turned north and the ascent became steeper. Ankole cattle herders walked along the sides of the road with staffs in their hands. Some men wore a woven blanket wrapped across one shoulder, and some wore broad-brimmed hats. These people were tall and thin, with narrower features and darker skins than the stocky Baganda around Kampala. Their Ankole cattle were unlike any I had seen before, with pointed horns, each about three to four feet long, that curved outward, then inward until they almost met.

Leaving them, we plodded up a mountainside where the land on one side dropped away to present a magnificent view across the extensive Maramagambo Forest and areas dotted with round, mirror-like lakes formed in old volcanic craters. This road was newly cut, exposing grayish-white volcanic ash

—283—

at the sides which were completely different from the red laterites to which we were accustomed. At one point the road was still under construction and we progressed upward along a dusty, boulder-strewn stretch where tractors and bulldozers were hard at work cutting through the rock. The car rattled, banged, and skidded on the thick ash, but kept going until we reached the top. Thinking the worst was over, we started to descend but came across a gang of workmen who diverted us onto a rough track running along at the side of the main road, which they had just prepared for tarmac. On one side of us was a tremendous drop into a green valley; in front, a huge yellow bulldozer plodded along very slowly.

We followed impatiently, going at a tedious pace for about half an hour until we reached a place where we were to join the main road again and where a pile of earth sloped up from the track to the tarmac. This was our undoing. We drove up to it, the car stuck, and then it stalled. A workman advanced towards us, gesticulating that we should reverse. We gesticulated back that it was impossible because the starter motor wouldn't work. This continued until we convinced him to come over and showed him the problem. He nodded and waved to a bunch of about twelve workmen, who obligingly came to help. Cliff steered, the children chattered on the back seat, and I got out while the men pushed the car, almost lifting it, through all obstacles onto the main road. After this they stood and waved cheerfully, but I saw their mouths drop open in amazement when I continued to push the car down the slope to get it rolling and leapt in once the engine turned over. I got covered in dust and grease, the car was full of dust, and we all looked like a pretty disreputable bunch.

Fortunately, we were almost there. Descending rapidly, we saw stretching ahead of us the plains surrounding the Queen Elizabeth National Park. We continued north, crossing the Kasinga Channel via the old Bailey bridge near Katunguru, where people fished from the banks of the river, before taking a sharp turn along a very rough road to reach the Katunguru gates of Queen Elizabeth Park just before closing time. After that, the road took us along the side of the Kasinga

THE ELUSIVE BABOON

Channel before we drove cautiously along the top of a narrow ridge that dropped to the Channel on one side and to the edge of Lake Edward on the other before the land opened up into the spoon-shaped Mweya Peninsula. There we found the lodge, parked, got out, and had a much needed wash. To our great pleasure, we discovered the Nuffield Lodge for scientists was completely different from Busingiro House for scientists. At Nuffield the amenities were more in line with my initial expectations of Busingiro. It had a dining-room-cum-sitting-room and eight guest bedrooms, each with running water, three flush toilets, and a room with a shower. An additional bonus was the food. This was cooked for us and brought to the table before fresh Uganda coffee was served.

While we waited for dinner, we saw wart-hogs, waterbuck, and bushbuck only a few yards from the window of the lodge at dusk. Then Erik came to say he had seen an elephant when he went out to look at the car. Thinking this was most unlikely and probably a product of a fertile imagination, which had led him to see dinosaurs in closets, we took little notice. But he insisted, and to humor him, I went with him to the door of the lodge. "Look, look!" he exclaimed triumphantly—and there was an elephant, just to the right of the door, tucking in to a bush in the garden. I grabbed Erik, pulled him inside, and leaned against a wall in shock at the thought of our son out there with that behemoth. I gathered my breath, stared at Erik, and said, "Don't you dare go out there again on your own."

Later that evening I asked about the elephant and was told it had been raised by someone in the vicinity. He had become tame when young and completely unafraid of people. Now he was full size and could be dangerous. Not only was he huge, he was often bad-tempered. Because of him, people were advised to use cars for safety if they moved even a few yards from the lodge, especially at night. My mind went back to the story we had been told in 1965 about the elephant that tossed the Germans' mini-van in the Queen Elizabeth Park. Maybe it was the same one. Meanwhile, I would have to watch Erik more closely.

CHAPTER 43

Beauty and Fear

CRANKING THE CAR ENGINE was a major drawback. We had either to rely on others to help push the car, or find a downward slope on which to park and roll forward to get it started. Maybe the problem wasn't difficult to fix but, unlike my dad, who loved cars, maintained his own, and would have sorted it out, Cliff had no practical interest in the workings of car engines. His only interest in cars was for transporting goods and getting from A to B. So after a good breakfast of bacon, eggs, tea, and toast, we headed to the mechanic on site, who worked on the tourists' buses and the scientists' Land Rovers and Jeeps, to ask him to fix it. We found him in overalls covered in black grease. Several vehicles stood on his lot. He frowned at our request before he rubbed his forehead with the back of his hand and told us he was busy all morning and unable to help until later. We had little choice but to stay on the peninsula and make sure we kept the engine running if we stopped on flat ground to look around.

First we visited the newly constructed Mweya Lodge, located on top of a rise about a hundred yards from the Nuffield Lodge. Mweya had been built for tourists and was fitted out like a first-class hotel. We parked facing downwards on the slope so we could push easily to start the engine, then got out and walked to the brow of the hill. Spectacular views opened up before us. To the north we could see the snow-capped peaks of the magnificent and legendary Ruwenzori Mountains, or "Mountains of the Moon." Below us, the blue waters of the Kazinga Channel, known for its rare birds, sparkled in the sun. When we saw those stunning views, our arduous journey the previous day seemed well worthwhile. We saw a ferry crossing the channel, and an assistant at the lodge said people could use it to visit the park on the other side. This cut the journey via the old iron Bailey bridge on which had we entered by about twenty-five miles. If the starter motor got fixed, we decided to take the ferry later.

Leaving the lodge, we drove steadily down to the side of the channel to visit an extensive hippopotamus wallow, where we parked, kept the engine running, and watched hundreds of animals luxuriating in the mud and water. Scarcely an inch was free from their large backs, which stuck out in huge leathery gray humps. Occasionally, one would lift its head and yawn to expose a cavernous mouth, and we saw two wrestle in a rather desultory manner, but for the most part they lay still, relaxed, enjoyed the warmth from the sun, and cooled in the thick and, to them, glorious mud.

When we returned in the evening they were beginning to stir and were making loud grunting, roaring, and wheezing noises. I thought some sounded like car engines revving up. One or two came to the bank, where they rooted around, while others trotted along the edge of the wallow, looking like huge pre-historic pigs, before they slumped back into the mud. The ground was heavily grazed and scattered with the skeletal remains of animals, making it possible to see how fossils formed near lakes, as had happened at Olduvai.

After lunch at the lodge, we went back to the mechanic. This time he greeted us, lifted up the hood, tinkered around, and in about ten minutes got the self-starter to work. We could therefore go further afield and decided to cross the channel by the ferry to explore the park on the other side. A vehicle track led down to the water's edge. We drove down slowly and came across three wiry, bare-footed young Africans in khaki shorts and white shirts who were talking and laughing. They smiled in welcome and prepared to ferry us across.

The ferry was an old-fashioned contraption with a short ramp leading to a flat wooden base. A metal rail ran along each side. On one side, a flat metal piece jutted out, and on this the driver sat at the steering wheel of the engine, which looked like a small tractor with paddles. Overhead, a small tin roof provided him with shade. We drove on, got out to look at the views and, as there were no other passengers, one of the men released a rope anchoring the ferry to a tree stump on the bank, let out a shout, jumped on board, and we were off. The ferry operated like a paddle steamer with limited power, but we moved forward at a steady pace, leaving a slight wake behind. Overhead, the sun shone from a clear blue sky and glinted on the glass-like surface of the water. We were in good spirits. The men laughed and chatted to one another.

Just over half way across, a startling change occurred. The wind began to blow, and choppy waves formed on the smooth surface of the water. To the east, the sky looked as though someone had drawn a horizontal line across with the side over us painted blue and the other side pitch-black. Menacing clouds like thick black smoke began to advance rapidly towards us. The blue quickly disappeared, and vast sheets of lighting rent the sky, followed by bone-shaking cracks of thunder that formed a cacophony of sound overhead. As the storm gathered strength, we had no protection from the rain that came down in torrents, nor did the driver's roof give him any protection. Waves began to flow over the bottom of the boat while the wind howled ferociously, catching at our clothes and hair, and pushing us off balance. It flattened our clothes against our skin, and soon we were soaked. The deck was too slippery for us to get to the car, and as the winds reached gale force I screamed that we must hold onto the rail and the children. By then the Kasinga Channel's current was so strong that if we were blown overboard we would be carried away in the roiling gray-green waters and drowned. Desperately hanging onto the rail and the children, we huddled together, praying the nightmare would soon end. Seeing our fear, the children clung to us, shivering and quiet.

Our driver hung onto the steering wheel, desperately trying to keep the boat on track. It struggled valiantly to move forward, but the wind was too much and began to blow us down the channel, causing the ferry to miss its mooring on the other side. The driver continued to spin the wheel to get the vehicle into a position where he could aim for the bank, but the limited power of the engine worked against him, and he was in a losing battle as he struggled against the raging storm. He began to pedal like mad, as if he were on a bicycle. Presumably he was working the paddles and trying to add extra power to the engine to move us forward, but what exactly he was doing was not clear. Water started to pour over the side at an alarming rate, and we feared the whole thing would sink.

Another African clung to the rail at the front and shouted orders to the steersman. The third hung on and kept lifting up a metal lid to look down into the bottom of the boat, as if to see whether water was coming in from underneath. The man at the wheel continued to steer frantically. And then, for a moment, the wind abated slightly, but it was enough for the driver to take advantage of the lull to get us closer to the bank. At this, the man in front took a flying leap over the water, landed on the shore, regained his balance and scrambled a few paces up the bank before turning back to face the boat. The second man threw him a long rope, which he grabbed and pulled on with all his might before looping it around a post to steady us. As the storm continued to pass over, the other men jumped to the bank, grabbed the rope, and the three of them heaved and pulled until they eventually managed to bring us to shore.

By then the ferry was turned back-to-front, but the wind had died down sufficiently for us to get into the car, and once docked, we reversed off the ferry and up the gentle slope of the bank onto firmer ground. We were soaked to the skin. Cliff seemed to recover quickly, but my legs felt like jelly. The children whimpered. While we debated what to do, another heavy storm broke. Thick puddles formed, and any vehicle tracks were converted into small streams, making progress impossible. There was no way we could explore the surrounding area, but we had to get back to the lodge.

After the hair-raising ride across the channel, I was reluctant to travel on the ferry again. But it was impossible to return via the bridge because the car would be brought to a standstill in the flooded land. With no alternative, we waited until the storm cleared and drove carefully down the muddy slope to where the ferry was moored. We were rewarded to see a gathering of majestic white pelicans, with black-tipped wings and large yellow bills, standing by the waters' edge. Apart from them, we saw little wild life.

We hailed the ferry and boarded. We were a forlorn little bunch with our wet clothes clinging to us. This time we went across with no further mishap and drove to the lodge to dry off, feeling we had achieved little that afternoon apart from a good soaking and being scared out of our wits. The following day we heard the ferry was out of action. It had sprung a leak!

CHAPTER 44

Frustration and Fun

THE NEXT DAY WE WENT IN SEARCH of chimpanzees. A researcher at the lodge had said they lived in a small forest near a guards' encampment and assured us they were "impossible" to miss. Because the ferry was not working, we took the long way round to the park. We'd just crossed the old Bailey bridge when we came across the rather repulsive sight of a mass of vultures fighting over, and tearing apart, a bloody carcass in a field at the side of the road with their sharp beaks. A trail of blood led to it, suggesting someone had hit the animal and then dragged it over there. We couldn't see what it was because the greedy vultures, with their tatty feathers spread out and bare necks burrowing into the corpse, completely blanketed the creature while turning it into a skeleton.

We drove on and soon came to the foot of the escarpment over which we had traveled two days earlier. There we saw a slight parting in the grass to our right. Cliff scrutinized a map and said this marked the track we were to follow. It looked as though no one had used it in ages. I was driving and refused to believe this was the way, but he insisted. We turned off the main road.

Tough grass had overgrown the track, making it barely discernible. On either side of us, the grass grew higher than the car and made it impossible to see exactly where we were going. Stones and hummocks were hidden in the grass, and we kept hitting them, causing me to fear the already frail car would fall apart. After a while, I refused to drive any further. Cliff thought I had not been trying hard enough. He took over only to find he was encountering the same difficulties, but he refused to admit defeat and kept driving. A Land Rover would have gone through easily, but we were in a small old sedan low to the ground. The children sat in the back, singing little songs, playing "I Spy," and asking when we would see the chimps.

The car plodded along, shaking us around as it rattled along the bumpy ground, and we seemed to be going nowhere. Even Cliff thought we should go back, but it was impossible to turn the vehicle around, and we were now too far in to back out. Suddenly we hit an ant-hill, causing me to be thrown up to the roof, where I hit my head hard. I yelled I was thoroughly sick of the whole thing before I sank into stony silence apart from muttering every so often that we should never have come. Cliff stared grimly ahead, tightened his lips, and said nothing.

After two hours of hostile tension, the track opened up, and we came to a game guards' camp. There, some men gaped in astonishment as we drove past waving. A grove of trees appeared. As they were the only trees around, we assumed they must be the ones we were looking for and stopped, but after the Budongo Forest, to which we were accustomed, these trees were very sparse. We waited about forty minutes but saw no sign of chimps and eventually gave up, pressed on, and soon found ourselves back at the main road from a totally different and much easier track. I refrained from pointing out this was the one we should have taken in the first place. So far we had spent the whole day pushing through tall grass, being bashed around, and seeing nothing.

Our next challenge was to get back to the lodge. Although the ferry wasn't working that morning, we thought it might have been fixed during the day, and as it was by far the shortest route, we turned off the main road to find out. The ferry was still out of action. We would have to go all the way round by the bridge and so we set off following a track at the side of the Kasinga Channel, feeling confident we would soon be back on the tarmac road—when, to our horror, the track petered out. We were lost.

Cliff got out of the car to see if he could figure out the way, and while he stood there, a herd of buffalo came towards him from behind. I yelled at him to get back in the car. He strolled back, opened the driver's door, and sat down, saying there was no need to make a fuss. If there was one animal it might charge, but that wouldn't happen with a herd. I was fed up and shut up. He started the engine, moved forward, and the buffalo moved off, but the light was beginning to fade, and we were running short of petrol. I'd had enough. My head still hurt from the bang it had received earlier in the day, and I had visions of being stuck all night in a car surrounded by buffalo and having to wait many hours before we were found.

At times like that, Cliff rose to the occasion and remained calm. I could see why he was an exceptional field person. While I sat and whined that we would never get out alive, he drove forward, trusting the petrol would last a bit longer even though the needle on the gauge pointed ominously to empty. Our luck held, and he eventually found a track, which in due time took us to the main road. Even if we stalled, we stood a much greater chance of getting help. All we had to do was find a petrol station. After about ten minutes, we spotted one and headed for it, heralding our arrival with a loud bang. The tire had blown again. We changed the wheel, filled up with gas, and drove off, reaching the park gates just before closing. Only then did the full impact of what had happened hit me. The mechanic in Nairobi had told us to treat the car gently, and we had ignored his warning. We had been more than fortunate that the car gave us no problems on the rough track in the long grass, or when we were off the main road at the side of the channel and surrounded by wild animals. We must have been mad to take so many risks in our unreliable vehicle with two small children in tow, and I wondered how two supposedly intelligent people could act like such idiots.

We'd had a difficult day and were thrilled when we encountered no more problems and got back to the tranquility of the lodge. Now we had the evening to look forward to, because we were going to Mweya Lodge to see some African dancers. We cleaned up, ate supper, visited the hippos again, and set off for Mweya, passing the tame, bad-tempered elephant along the way. A crowd of white tourists and some African onlookers had gathered on the flood-lit lawn to watch the dancers, who were led by a small thin man with amazing stomach muscles that looked like a wash-board. He rapidly pushed them in and out then wiggled his body so the muscles rippled upwards from the bottom towards his head. The European visitors took this amazing feat very seriously, but the Africans who sat on a bench to watch kept going into fits of laughter. At one point the dancing man went over to a white woman and wiggled in front of her at which the African laughter increased. One man laughed so hard he fell off the bench making me wonder whether there was some inside joke and whether the dancer was mocking the tourists. None of the other dancers behaved in such an odd manner but performed graceful movements to the rhythmic sound of drums, and the Africans onlookers stayed calmly on the bench.

After our harrowing day, the dancers were a great treat, and once the performance was over, we returned to the lodge to relax in the lounge and talk with other visitors. At around ten we heard lions roaring. I had never seen them in the wild, and when one of the researchers said he knew where to find them and would take us in his Land Rover to look at them, we were excited to go. Our children were up late, so we climbed into the vehicle, joined by one or two other people from the lodge who squeezed in. The driver gunned the engine, and we were off. For the next half hour we sped around the peninsula at a terrifying speed. There was hardly any moonlight and only the car headlights to show the way. Clinging on tightly, we were headed towards the

THE ELUSIVE BABOON

precipitous edge of the peninsula before the driver swung the wheel sharply at the last minute. This caused several gasps and a few nervous titters. Some brave soul plucked up the courage to ask if everything was alright. Our cheerful driver told us not to worry; he was used to the area. He continued to speed along the track.

Again, the headlights picked out the landscape, and we headed to the very edge of the peninsula. Once more the driver swung the wheel at the last minute. But we'd seen steep drops at the side of the road and prayed he would not take a wrong turn and plunge us into the dark waters below.

Needless to say, we saw no lions. Meanwhile, apart from Cliff, who as usual seemed unfazed, our happy little group of passengers had turned into nervous wrecks whose sole aim was to get back to the safety of the lodge. Eventually the exhilarating ride was over. We pulled up, got out, congratulated one another on our survival, and stumbled into the lodge for a stiff drink. Thus we celebrated the end of strange day. I wondered what excitement awaited us tomorrow.

CHAPTER 45

The Long Trek Back

EARLY THE NEXT MORNING, we packed the car, ate breakfast, said goodbye to our hosts at the lodge, and set off for Kampala in brilliant sunshine, planning to be there by late afternoon. We retraced our steps to the escarpment and got another flat tire. But by then we were experts and changed the wheel in ten minutes before moving slowly forward in second gear up the winding road that led over the escarpment. Near the top, a gang of men was working on the road, and a bulldozer was pushing down rocks to lower a hill. Rocks were strewn in our path, and we had to wait for them to be cleared but were making good time. The miles began to roll away, and when we reached the tarmac on the other side, we were in fine spirits.

Suddenly there was an enormous bang, and the car swerved in the road. We had had another blow-out. A little voice from the back squeaked, "Oh, no, not again," as I drew to a halt and switched off the engine. Cliff said we should pull off the road. I turned the starter key. The only response was a click. I threw up my hands in frustration. The newly repaired starter-motor had gone again. We were back to pushing.

THE ELUSIVE BABOON

But worse than this, realized we could go no further without having the tire mended, for even the spare had a hole. Our only alternative was for Cliff to hitch a lift to a petrol station and try to get one of the tires repaired. We got out, shoved the car to the roadside, and I swore to myself I would never again travel in such an unreliable, anxiety-provoking vehicle. The journey was getting more and more ridiculous. For the children, it had become a game in which they took bets on when another tire would blow.

Very few cars traveled that way, and it was getting hot as we sat on the hard ground at the roadside, hoping someone would come by before too long. Thirty minutes passed before a small gray Renault came into view. It appeared to be full of people. This was no surprise, because full cars were standard practice in Uganda, but we were desperate and waved frantically. Maybe they could help. The car slowed down and came to a stop. A rather fat bearded *mzungu* jumped out, hitched up his pants, and came over to see what was going on. We explained our predicament while he listened carefully. He saw me with the two children and then cheerfully said he could give Cliff a lift to the next petrol station even though his car was already pretty full. We were thrilled. Cliff picked up the spare wheel and somehow wedged in before they drove off in a cloud of dust.

While the children and I sat by the roadside, Africans came to talk to us quite unselfconsciously in their own language and it didn't seem to matter that we couldn't talk back. A group of children made faces at our children as a way to communicate, and I was glad of their company. It helped to pass the time, so it didn't seem too long before Cliff came back in a minibus that had picked him up at the petrol station where he'd had the tire repaired. Surrounded by a crowd of curious onlookers, we changed the wheel again, and having received a push from several locals who were highly amused by this activity, the engine turned over, and we rolled into action and waved goodbye.

Moving at a steady pace we stopped along the way at each petrol station to buy a new inner tube but with no success. If we were unable to find one in

Mbarara, we were out of luck. In Mbarara, the shops were closed because it was Sunday. A petrol station sold inner tubes, but none fit our vehicle. The attendants at another station thought they might have the right tube, but their manager had the keys to the storage room, and he was not available on Sunday.

The last thing we wanted was to stay overnight in Mbarara on the off-chance we could buy an inner tube on Monday. We had heard a rumor that two Americans, a reporter and sociologist, who were poking into army business, had disappeared from there not long before and were presumed killed. We were doing nothing suspicious, but in the uncertain environment that prevailed, who knew what people might think of us and what might happen?

We decided to press on, but after twelve miles the tire blew again, leaving us with two alternatives. Either we drove back to Mbarara or we risked a seventy-mile journey to Masaka, across a deserted plain where we could be stranded without any help if we ran into more trouble. Tired and dispirited, we changed the wheel and drove back. To add to our difficulties, the car now juddered violently when we reached thirty miles an hour, and it was difficult to hold the steering wheel. We had changed the wheels so often they were out of line. But we had no more blowouts. In Mbarara we found a suitable hotel, and the children ran around happily outside. I was glad Cliff took things in stride, because the strain was taking its toll on me. We were stuck in a foreign country with a dodgy vehicle, were miles from any reliable garage, had a time deadline to catch a plane, many miles of driving ahead, and two small children in tow. Instead of reaching Kampala in the afternoon as planned, we were unexpectedly delayed for another day.

A good night's sleep helped. The following morning we tracked down a resourceful Indian mechanic who, to our intense relief, found a new inner tube and fitted it. He couldn't fully align the wheels, so our speed was limited to about thirty miles an hour, but we could move forward. Citroens, Volkswagens, Volvos, and Peugeots zipped past at about seventy while we moved along at a tortoise-like pace, praying the car would keep going. Gradually the miles rolled by, and we finally limped into Kampala in mid-afternoon, where our first order of business was to find a reliable garage. The car required a complete overhaul before we set off on the long trip back to Nairobi. Perhaps it was a good thing that we'd had to delay our journey and had experienced problems on our trip to the park rather than when we had to go over the Kenya Highlands. To have had blow-outs and no starter motor on that journey could have been a disaster. This time we had to find someone we could trust to do the work properly.

John Goodman, our old friend from 1966, was still in Uganda and recommended Fergie Wharton's Garage in the industrial area. Placing our trust in John, we took the invalid to Fergie, hoping he could work magic on it. The car stayed there for the next day while we got our things together, took the children to the swimming pool at Makerere, and I visited the Nommo Art Gallery, where I purchased a batik depiction of a village scene by the Ugandan artist Mark Mutyaba. The Goodmans still lived on Katalemwa and were getting ready to return to the States, but generous as always, they asked us out there for a meal, and we saw our old house. To our surprise and pleasure, we saw Lawrence, who as usual was full of doom and gloom like Eeyore. We were unable to find out what had happened to Fred, Robin, and Jimmy.

The next day we collected the dehydrated food samples and at the end of the day picked up the car. The repairs were expensive, but we were assured it was in good shape and ready to travel the long four hundred and eight miles to Nairobi. It was impossible for our car to do that in one day, but if it gave no more trouble, we would reach Nairobi just in time to get rid of it and do some quick shopping before leaving for the airport. The leisurely trip we had planned from Kampala to Nairobi was out of the question.

We set off early the following day with a full load that included the dehydrated samples. Driving against the clock we took turns at the wheel in a car that was still unable to go more than forty-five miles an hour. We drove through warm, sunny plains to Tororo, went over the border, then started to climb up into the Kenya Highlands to a height of about nine thousand feet, hitting torrential rain, low cloud cover, and freezing sleet on the way to Eldoret.

I was driving when we ran into a blizzard. Visibility was poor, and the windshield wipers swished slowly backwards and forwards as I peered ahead, but the car valiantly plodded on. Fergie had done a good job.

We came across snow at the side of the road near the town of Equator, and it was extremely cold. My shoulders were tense as I hunched over the wheel, and I had to concentrate hard but kept going. There was no other traffic. All we saw were some people rushing along with sacks over their heads and covering their bodies to the knees to protect them against the rain. Wisps of smoke rose from some huts in a stretch of pine forest.

At about four o'clock we reached Eldoret. With time so much against us, we decided to press on although it was another ninety miles to Nakuru. The rain continued but the road became easier as we dropped into the Rift Valley, moved onto flatter ground and reached Nakuru just before dark. After searching around, we found a hotel, settled in, and inspected the car. One of the tires was almost flat, but it had seen us safely over the Highlands. The following day we bought another new inner tube and paid a quick trip to Nakuru National Park. Again we had magnificent views of exotic long-necked flamingos in such abundance that their feathers formed pink and red bands across the lake. There were long-beaked pelicans, marabou storks, waders, cormorants, and a variety of ducks. We wanted to spend more time but had to keep going and continued at a steady pace through the Rift Valley, crossing the railway in places and even seeing smoke from the train as it toiled up a ridge in the distance. We kept our promise and bought some sheepskins rugs when we were entering Nairobi, and late that afternoon headed to our old Ainsworth Hotel, where we finally relaxed, feeling greatly indebted to our dear friend John Goodman. He had recommended an excellent mechanic in Fergie Wharton, and his good services had seen us through. We had one day left in which to sell the car, buy presents, and pack.

CHAPTER 46

Final Days

WITH NO TIME TO PUT AN AD IN A LOCAL PAPER, we tried to sell the car the following day by visiting rows of showrooms full of new and second-hand cars on one of the main streets where Cliff had purchased our vehicle. When he tried to sell it back to the dealer who sold it to him, the man said, "Oh, yes, I remember it. Unfortunately, business is not good at the moment, and I have no room for it."

"I bet he remembers it," I muttered as I thought back to the garage mechanic's comment that it was held together with wire and rubber bands. We told him how much we'd improved the car and that if we were getting a bargain in the first place, as he had claimed, he would be getting a much better bargain by buying it back at a reduced rate. He was not persuaded, and so we trailed wearily from one dealer to another. Only one was interested. He insisted we show him how the car worked, and we ended up driving him all over Nairobi. After this, he made an insultingly low offer that was no more than we had paid to have it fixed in Kampala. We refused because, even with its limitations, we knew the car was worth more.

But as we continued to trail around getting nowhere and began to run out of time, we were left with no choice but to accept the only offer. I felt resentful when we handed over the car to the dealer with his greased hair, dapper light-gray suit, and shiny black shoes. Still, it was a relief to be free of that worrisome old vehicle. An hour or two remained to buy souvenirs.

Nairobi was full of gift shops containing many items created just for tourists. We knew from experience that some dealers were unreliable because we had a friend in New York who had paid for goods to be sent to the States and never received them. Nor had he received any response to the many letters he sent asking what had happened. We went to a place highly recommended by local people. The outside was not impressive, with a smallish window on each side of a wooden door, but we entered and found ourselves in a large space packed with goods. Many were authentic pieces that were piled on shelves lining the room or packed in baskets all over the floor.

After our eyes adjusted to the dim light, we saw, to the right of the entrance, an enormous basket full of fearsome looking spears that had belonged to Masai warriors and were quite genuine. There were some imitations in other parts of the shop, but they were selling for much less. We were tempted to buy one, but there was no way to transport a spear back on the airplane and through Customs at Kennedy Airport. Round-topped wooden stools with three legs curving outwards for balance were crowded together next to the spears. Some tables were like tiny toys; others had surfaces the size of coffee tables.

Drums had tops wider than their bases. Zebra or antelope skin stretched across the surface. Carved wooden baboons, elephants, warthogs, gazelles, giraffes, hippos, and human figures had been made specifically for tourists. There were musical instruments; brightly colored necklaces of beads in greens, reds, yellows, blues, and white, and some made of cowrie shells; and all kinds of gourds for household use. A smell of incense tinged with the aroma of grease and smoke rose from the genuine artifacts. More recent ornaments were not so valuable but were interesting because they incorporated modern products like the cap from a Coca-Cola bottle.

Towards the back of the shop a wooden staircase, built like a ladder with broad flat rungs, led to a loft. We climbed up and found ourselves in a room that smelled of the same grease and smoke. A series of shields hung from the ceiling. One was plain brown, enormous, and expensive. To me it looked tattered and battered, but Cliff became excited and would have bought it if he could have afforded it. He said it was an authentic piece that was old and quite rare because shields like it were no longer made. Constructed of buffalo hide, it was thick and heavy and must have required an exceptionally strong warrior to carry it. Ideally, it should have been in a museum and not in a shop or someone's home. Pots and containers made from gourds, hides, or horns were stacked on the shelves. While we pondered over these treasures, the European owner with a tanned, weather-beaten face stopped by to talk. He told us he journeyed into the country every few weeks to acquire genuine pieces, but they were becoming more and more difficult to find.

And now our time was up, and we needed to choose. We ended up with a motley collection. This included a Masai woman's necklace made of rows and rows of small red, green, and blue beads strung on several rows of wire and held in position by strips of leather arranged at intervals in a radial pattern to form a collar. They smelled of smoke and were covered in red ocher. Another purchase was an enormous necklace of ostrich-shell beads with each bead cut into a round shape about half an inch in diameter with a hole pierced in the middle. We bought a young girl's pubic apron, consisting of a triangular-shaped piece of leather bounded by rows and rows of ostrich shell beads. This would be placed over the private parts and held in place by being suspended from around the hips using leather strips just above the buttocks. It had been worn, and this added to its value. We purchased a belt made with red and green beads and dik-dik bones. In addition, we bought a leather band to go around the ankle. Ornaments cut out of gourds and shaped like bells, which clicked and clacked as you moved, dangled from it.

We picked out some gourds decorated with cowrie shells and colored beads. One we particularly liked had been used for milk and had a small piece of elephant-tail hair woven into it, which pushed up the price considerably. The shop's owner told us this was because the hair was getting difficult to come by, although there were many imitations of the real thing. A particularly authentic-looking plastic was now on the market and used to make elephant-hair bracelets, which were very popular at that time.

Our final purchase was a lethal-looking wrist knife made of iron that was curved to go over the wrist and about an inch wide. Lined with protective leather on the inside and outside, it had a sharp outer edge. The wearer would put it onto his wrist, take off the leather on the outside and slash at the enemy in a scuffle. We gathered everything together, paid, and left our purchases to be sent to New York.

It was time to head to the airport. Cliff was flying to Ethiopia to prepare for his next field trip, where his team later trapped, sedated, bled, and released many baboons based on what he learned during our first trip to Uganda. Meanwhile, I waited with the children to board a flight to London to stay with relatives. After a few days, we all met up before returning to Manhattan. Our parcel of goods from Nairobi arrived some months later. When opened, it exuded the smell that pervaded the shop and reminded us of the pleasant time we had spent there.

My Ugandan journey was over. I was back with the skyscrapers and my memories. A siren screamed, tires screeched, the air-conditioner whirred.

ABOUT THE AUTHOR

Jennifer Jolly grew up in North Lincolnshire, England, and moved to London in 1957 and to the United States in 1967. For most of her career she consulted with major U.S. corporations on the human resource implications of business initiatives, and on the identification and development of future leaders. She has a B.Sc. from London University and a Ph.D. in Industrial/Organizational Psychology from New York University. In 1979 she received the Douglas H. Fryer award for academic excellence and research on Behavior in Organizations. She is the author of *Job Change: Its Relationship to Role Stresses and Stress Symptoms According to Personality and Environment* and is a co-author of *Teams: Who Needs them and Why*. Now retired, she lives with her husband, the physical anthropologist Clifford J. Jolly, in Hoboken, New Jersey.

www.ingramcontent.com/pod-product-compliance
Lightning Source LLC
Chambersburg PA
CBHW031313160426
43196CB00007B/520